Hen's Teeth and Horse's Toes

Stephen Jay Gould

PENGUIN BOOKS

PENGUIN BOOKS

Published by the Penguin Group
27 Wrights Lane, London W8 5TZ, England
Viking Penguin Inc., 40 West 23rd Street, New York, New York 10010, USA
Penguin Books Australia Ltd, Ringwood, Victoria, Australia
Penguin Books Canada Ltd, 2801 John Street, Markham, Ontario, Canada L3R 1B4
Penguin Books (NZ) Ltd, 182–190 Wairau Road, Auckland 10, New Zealand

Penguin Books Ltd, Registered Offices: Harmondsworth, Middlesex, England

First published in the USA by W. W. Norton & Company 1983
Published in Pelican Books 1984
Reprinted in Penguin Books 1990
10 9 8 7 6 5 4 3 2 1

Made and printed in Great Britain by
Richard Clay Ltd, Bungay, Suffolk

Penguin Books
Hen's Teeth and Horse's Toes

Stephen Jay Gould grew up in New York City. He graduated from Antioch College and received his Ph.D. from Columbia University in 1967. Since then he has been Professor of Geology and Zoology at Harvard University. He considers himself primarily a palaeontologist and an evolutionary biologist, though he teaches geology and the history of science as well. A frequent and popular speaker on the sciences, his published work includes *Ontogeny and Phylogeny*, a scholarly study of the theory of recapitulation; *The Mismeasure of Man* (Penguin 1983), winner of the National Books Critics' Circle Award for 1982; the popular collections of essays *Ever Since Darwin: Reflections in Natural History* (Penguin 1980), which received great acclaim: 'Unreservedly, they are brilliant' – *New Scientist; The Panda's Thumb: More Reflections in Natural History* (Penguin 1983), which won the 1981 American Book Award for Science; *The Flamingo's Smile* (Penguin 1987), *Time's Arrow, Time's Cycle* (Penguin 1988) and *An Urchin in the Storm* (Penguin 1989).

For My Mother

Brave woman
Wise owl

Contents

Hen's Teeth and Horse's Toes

Prologue

ALL THE WORLD LOVES a centennial; we can't resist the temptation to celebrate something clean and even in a ragged and uncertain world. I am gathering this third volume of collected essays* in the midst of worldwide festivities for the third Darwinian centennial of our century. The first, in 1909, celebrated the 100th anniversary of Darwin's birth; the second, in 1959, the 100th year since publication of the *Origin of Species;* the third, in 1982, the 100th anniversary of his death. Darwin and evolutionary theory have been the focal point of all my writing in this series (my personal tribute to Darwin on his third centennial appears as essay 9 in this volume). The sequence of centennials provides us with a fine epitome of evolutionary theory in our century, and with some insight into its present successes and tribulations.

The organizers of the major 1909 shindig at Cambridge University had to cover up an embarrassing fact in preparing their hagiographical centennial volume. Although no

---

*The first two volumes, *Ever Since Darwin* and *The Panda's Thumb,* were published by W.W. Norton in 1977 and 1980 and are available in both hard and paperback editions. Again, the source of nearly all essays is my monthly column in *Natural History Magazine,* entitled *This View of Life.* Three essays did not originate there. Number 19 first appeared in *Discover Magazine* for May 1981; essay 24 was written for *Junk Food* published by the Dial Press, 1980. I wrote essay 17 for this volume as a commentary upon criticism that essay 16 has received since it first appeared in *Natural History.*

thinking person doubted the fact of evolution by 1909, Darwin's own theory about its mechanism—natural selection—was not then at the height of its popularity. Indeed, 1909 marked the acme of confusion about how evolution happened in the midst of complete confidence that it had occurred. An embattled group of strict Darwinians, led by the aging A. R. Wallace in England and by A. Weismann in Germany, continued to hold that virtually all evolutionary change occurred by the cumulative power of natural selection building adaptation step-by-step from the random raw material of small-scale genetic variation. Lamarckism remained strong and provided an alternative to natural selection for the gradual building of adaptations—creative organic response to perceived needs and the transmission of these favorable responses to offspring through the inheritance of acquired characters. Mendelian inheritance, when properly elucidated, tipped the scales in Darwin's favor; but, in 1909, it had (in the gropings of its youth) merely sown more confusion by adding yet a third mechanism to the swirling competition—production of new species all at once by large and fortuitous mutations.

By 1959, confusion had ceded to the opposite undesired state of complacency. Strict Darwinism had triumphed. The flowering of Mendelian genetics had finally laid Lamarckism to rest since the workings of DNA provided no mechanism for an inheritance of acquired characters. Early fascination with large mutations had ceded to a recognition that copious and continuous small-scale variation also had a Mendelian base, and provided a far better source for the raw material of evolutionary change than occasional and detrimental large mutations. But random, small-scale variation produces no change by itself and requires a shaping force to preserve and enhance its favorable component. By 1959, nearly all evolutionary biologists had concluded that natural selection, after all, provided this creative mechanism of evolutionary change. At age 150, Darwin had triumphed. Yet, in the flush of victory, his latter-day disciples devised a version of his theory far narrower than anything Darwin himself would have allowed.

The strict version went well beyond a simple assertion that natural selection is a predominant mechanism of evolutionary change (a proposition that I do not challenge). It emphasized a program for research that almost dissolved the organism into an amalgam of parts, each made as perfect as possible by the slow but relentless force of natural selection. This "adaptationist program" downplayed the ancient truth that organisms are integrated entities with pathways of development constrained by inheritance—not pieces of putty that selective forces of environment can push in any adaptive direction. The strict version, with its emphasis on copious, minute, random variation molded with excruciating but persistent slowness by natural selection, also implied that all events of large-scale evolution (macroevolution) were the gradual, accumulated product of innumerable steps, each a minute adaptation to changing conditions within a local population. This "extrapolationist" theory denied any independence to macroevolution and interpreted all large-scale evolutionary events (origin of basic designs, long-term trends, patterns of extinction and faunal turnover) as slowly accumulated microevolution (the study of small-scale changes within species). Finally, proponents of the strict version sought the source of all change in adaptive struggles among individual organisms, thus denying direct causal status to other levels in the rich hierarchy of nature with its "individuals" both below the rung of organisms (genes, for example) and above (species, for example). The strict version, in short, emphasized gradual, adaptive change produced by natural selection acting exclusively upon organisms.

At the second centennial, some experts even declared that the immense complexity of evolution had yielded to final resolution. One leader remarked in a famous essay: "Differences of opinion on relatively minor points naturally persist and many details remain to be filled in, but the essentials of the explanation of the history of life have probably now been achieved."

Now, at the third centennial, Darwinian theory is in a vibrantly healthy state. Confidence in the basic mechanism of natural selection provides a theoretical underpinning and point of basic agreement that carries us beyond the pessimistic anarchy of 1909. But the constraints of an overzealous strict version, so popular in 1959, are loosening. Exciting discoveries in molecular biology and in the study of embryological development have reemphasized the integrity of organic form and hinted at modes of change different from the cumulative, gradual alteration emphasized by strict Darwinians. Direct study of fossil sequences has also challenged gradualistic biases (the "punctuated equilibrium" pattern of long-term stasis within species and geologically rapid origin of new species) and asserted the idea of explanatory hierarchy in identifying species as discrete and active evolutionary agents (just as molecular biology has extended hierarchy in the other direction by discovering evolutionary processes at genic levels that are "invisible" to organisms—see essay 13).

However, and ironically, the early 1980s have also witnessed an utterly different and perverse debate about evolution, often conflated in the public mind with these legitimate and exciting arguments about evolutionary mechanisms. I refer, of course, to the political resurgence of the pseudoscience known to its supporters as "scientific creationism"—strict Genesis literalism masquerading as science in a cynical attempt to bypass the First Amendment and win legislatively mandated inclusion of particular (and minority) religious views into public school curricula. As in 1909, no scientist or thinking person doubts the basic fact that life evolves. Intense debates about *how* evolution occurs display science at its most exciting, but provide no solace (only phony ammunition by willful distortion) to strict fundamentalists.

This peculiar juxtaposition of utterly different debates ostensibly about the same subject reminds me either of Wagner's two operas about song contests, *Tannhäuser* and *Die Meistersinger*—one sublime, the other comic; or of Spielberg's two films about oddities in suburbia for a 1982 summer, *E.T.* and *Poltergeist*—one joyful, the other sinister. But life is a continual conflation of sacred and profane, and who would have it any other way?

The present collection of essays about evolution was born amidst these tensions. It treats both the purely political and non-intellectual controversy stirred up by modern creationists (Section 5) and the fascinating debates now occurring within evolutionary theory. I discuss, for example, the role of alterations in embryonic development as a possible mechanism for rapid evolutionary transformations (Section 3); chance as a source of evolutionary change, not merely raw material (essay 26); evolution at hierarchical levels both above and below the traditional Darwinian focus on organisms (essay 13); and constraints of development and inheritance as arguments for the integrity of organisms and against an overly atomistic and deterministic view of adaptation (Section 3, but also as a major theme in essays 3, 10, and 29).

An ultimately more important tension—one that makes evolution such an exciting subject both for scientists and for all of us—contrasts these lively controversies that divide us with the enormous uniting power and extension of evolutionary theory itself. As I have explored the arguments, so too have I written about a view of life that has, since Darwin's day, transformed our concept of ourselves and our world. "Big" questions about the history of our earth and its life provide a way to explore the thought of exemplary past scientists, and to understand the process of science itself, when done best (Section 2). (If readers emerge from this section with the message that science traffics in answerable questions, not in every fascinating reverie of the human mind, they will also understand why modern creationism is not science.) The seepage of evolutionary questions into ostensibly distant political debates (Section 5) displays both the pervasive scope of this view of life and also the inextricable union of scientific and social issues. The broader and distinctive view that emerges from evolutionary theory could enlarge our concept of science and of explanation in general by emphasizing historical contingency and quirky change (but sensible in retrospect) against the predictable and regular world preached by the stereotype of so-called "hard" science. I explore this issue throughout, but primarily in essay 4 (the one piece that should probably be read twice if you deem any essay worthy of such attention).

These issues are all abstract; but my vehicle for raising them remains the peculiar and mysterious particulars of nature. I have never been able to raise much personal enthusiasm for disembodied theory. Thus, when I wish to explore the explanatory power of evolutionary theory (Section 1), I write about apparent oddities resolved by Darwin's view—dwarf male anglerfishes parasitically united with females, wasps that paralyze insects to provide a living feast for their larvae, young birds that kill their siblings by simply pushing them outside a ring of guano serving as a "nest," and male mites that cycle through their lives in a fraction of the time alotted to females. Other essays treat generalities through the agency of particular mysteries; why are the genitalia of female spotted hyenas dead ringers for the male penis and scrotum; why do no large animals move on wheels; how can hens be induced to form teeth when no birds have produced them for more than fifty million years; how can some flies form legs in their mouths (I even discuss an "ouchless" mosquito, so afflicted); why did the demise of dinosaurs coincide with an extinction of at least twenty-five percent of marine invertebrate families; are zebras white on black or black on white, and what general rule unites their varied patterns of striping? I even think that a generality lies behind my piece on shrinking Hershey Bars, but I won't push it. Pure humor (or attempts thereat) has its place as well.

Darwin, at his third centennial, would be satisfied indeed with the vigor of his child, now grown so large and strong. He would also welcome the legitimate and far-reaching debates that surround his theory, for absence of dogmatism is the truest mark of a great scientist. At the first centennial of 1909, William Bateson (see essay 11), perhaps the least Darwinian of all participants, paid the finest of all tributes by writing of Darwin: "We shall honor most in him not the rounded merit of finite accomplishment, but the creative power by which he inaugurated a line of discovery endless in variety and extension."

1 Sensible Oddities

1 | Big Fish, Little Fish

ALFRED, LORD TENNYSON, never known for egalitarian perspectives, had this to say about the relative merit of the sexes:

> Woman is the lesser man, and all
> thy passions, matched with mine,
> Are as moonlight unto sunlight, and
> as water unto wine.

The couplet may not represent Tennyson's considered view, since the protagonist of "Locksley Hall" had lost his love to another and speaks these words during a grand poetic fit of sour grapes. Still, the literal reading—that women are smaller than men—would be accepted by most of us as a general fact of nature, not as a sexist trap. And most of us would therefore be wrong.

Human males are, of course, generally larger than human females, and most familiar mammals follow the same pattern (but see essay 11). Yet females are larger than males in a majority of animal species—and probably a large majority at that. For starters, most animal species are insects and female insects usually exceed their males in size. Why are males generally smaller?

One amusing suggestion was proposed in all seriousness just 100 years ago (as I discovered in the "50 and 100 Years Ago" column in *Scientific American* for January, 1982). A certain M. G. Delaunay argued that human races might be ranked by the relative social position of females. Inferior races suffered under female supremacy, males dominated in superior races, while equality of sexes marked races of middle rank. As collateral support for his peculiar thesis, Delaunay argued that females are larger than males in "lower" animals and smaller in "higher" creatures. Thus, the greater number of species with larger females posed no threat to a general notion of male superiority. After all, many serve and few rule.

Delaunay's argument is almost too precious to disturb with refutation, but it's probably worth mentioning that the paradigm case of a "higher" group with larger males—the mammals—is shakier than most people think (see Katherine Ralls in the bibliography). Males are larger in a majority of mammalian species, of course, but Ralls found a surprising number of species with larger females, spread widely throughout the range of mammalian diversity. Twelve of 20 orders and 20 of 122 families contain species with larger females. In some important groups, larger females are the rule: rabbits and hares, a family of bats, three families of baleen whales, a major group of seals, and two tribes of antelopes. Ralls further reminds us that since blue whales are the largest animals that have ever lived, and since females surpass males in baleen whales, the largest individual animal of all time is undoubtedly a female. The biggest reliably measured whale was 93.5 feet long and a female.

The sporadic distribution of larger females within the taxonomic range of mammals illustrates the most important general conclusion we can reach about the relative size of sexes: the observed pattern does not suggest any general or overarching trend associating predominance of either sex with anatomical complexity, geological age, or supposed evolutionary stage. Rather, the relative size of sexes seems to reflect an evolved strategy for each particular circumstance—an affirmation of Darwin's vision that evolution is primarily the story of adaptation to local environments. In this perspective, we must anticipate the usual pattern of larger females. Females, as producers of eggs, are usually more active than males in brooding their young. (Such male tenders as sea horses and various mouth-brooding fishes must receive eggs directly from a female or actively pick up eggs after a female discharges them.) Even in species that furnish no parental care, eggs must be provided with nutriment, while sperm is little more than naked DNA with a delivery system. Larger eggs require more room and a bigger body to produce them.

If females provide the essential nutriment for embryonic or larval growth, we might ask why males exist at all. Why bother with sex if one parent can supply the essential provisioning? The answer to this old dilemma seems to lie in the nature of Darwin's world. If natural selection propels evolution by preserving favored variants from a spectrum randomly distributed about an average value, then an absence of variation derails the process—for natural selection makes nothing directly and can only choose among alternatives presented. If all offspring were the xeroxed copies of a single parent, they would present no genetic variation (except for rare new mutations) and selection could not operate effectively. Sex generates an enormous array of variation by mixing the genetic material of two creatures in each offspring. If only for this reason, we shall have males to kick around for some time.

But if the biological function of males does not extend beyond the contribution of some essentially naked DNA, why bother to put so much effort into making them? Why should they, in most cases, be almost as big as females, endowed with complex organs, and quite capable of an independent life? Why should industrious bees continue to make the large and largely useless male creatures appropriately known as drones?

These questions would be difficult to answer if evolution worked for the good of species or larger groups. But Darwin's theory of natural selection holds that evolution is fundamentally a struggle among individual organisms to pass more of their genes into future generations. Since males are essential (as argued above), they become evolutionary agents in their own right; they are not designed for the benefit of their species. As independent agents, they join the struggle in their own ways—and these ways sometimes favor a larger size. In many groups, males fight (literally) for access to females, and heavyweights often have an edge. In more complex creatures, social life may emerge and become ever more elaborate. Such complexity may require the presence and active involvement of more than one parent in the rearing of offspring—and males gain a biological role transcending mere stud service.

But what of ecological situations that neither favor battle nor require parental care? After all, Tennyson's most famous biological line—his description of life's ecology as "Nature, red in tooth and claw"—does not apply in all, or most, cases. Darwin's "struggle for existence" is a metaphor and need not imply active combat. The struggle for genetic representation in the next generation can be pursued in a variety of ways. One common strategy mimics the motto of rigged elections: vote early and vote often (but substitute "fornicate" for "vote"). Males who follow this tactic have no evolutionary rationale for large size and complexity beyond what they need to locate a female as quickly as possible and to stick around. In such cases, we might expect to find males in their minimal state, a status that might have become quite general if evolution worked for the good of species—a small device dedicated to the delivery of sperm. Nature, ever obliging, has provided us with some examples of what, but for the grace of natural selection, might have been my fate.

Consider a species so thinly spread over such a broad area that males will rarely meet at the site of a female. Suppose also that females, as adults, move very little if at all: they may be attached to the substrate (barnacles, for example); they may live parasitically, within another creature; or they may feed by waiting and luring rather than by pursuit. And suppose finally that the surrounding medium can easily move small creatures about—as in the sea, with its currents and high density (see M. Ghiselin's book, *The Economy of Nature and the Evolution of Sex,* for a discussion of this phenomenon). Since males have little impetus for literal battle, since they must find a stationary female, and since the medium in which they live can provide (or substantially aid) their transport, why be large? Why not find a female fast when still quite small and young and then hang on as a simple source of sperm? Why work and feed, and grow large and complex? Why not exploit the feeding female? All her offspring will still be 50 percent you.

Indeed, this strategy is quite common, although little appreciated by sentient mammals of different status, among marine invertebrates that either live at great depth (where food is scarce and populations very thinly spread), or place themselves in widely dispersed spots that are hard to locate (as in many parasites). Here we often encounter that ultimate in the expression of nature's more common tendency —females larger than males. The males become dwarfs, often less than one-tenth the length of females, and evolve a body suited primarily for finding females—a sperm delivery system of sorts.

A species of *Enteroxenos,* for example, a molluscan parasite that lives inside the gut of sea cucumbers (echinoderms related to sea urchins and starfishes), was originally described as a hermaphrodite, with both male and female organs. But J. Lutzen of the University of Copenhagen recently discovered that the male "organ" is actually the degenerated product of a separate dwarf male organism that found the parasitic female and attached permanently to her. The female *Enteroxenos* fastens herself to the sea cucumber's esophagus by a small ciliated tube. The dwarf male finds the tube, enters the female's body, attaches to it in a particular place, and then loses virtually all its organs except, of course, for the testes. After a male enters, the female breaks its tubular connection with the sea cucumber's esophagus, thereby obliterating the pathway of entrance for any future males. (A strict Darwinian—I am not one—would predict that the male has evolved some device to break or cause the female to break this tubular connection, thereby excluding all subsequent males and assuring its own paternity for all the female's offspring. But no evidence yet exists for or against this hypothesis.)

As long as such an uncomfortable phenomenon resides with unfamiliar and "lowly" invertebrates, male supremacists who seek pseudosupport from nature may not be greatly disturbed. But I am delighted to tell a similar story about one group of eminently suited vertebrates—deep-sea anglerfishes of the Ceratioidei (a large group with 11 families and nearly 100 species).

Ceratioid anglerfishes have all the prerequisites for evolving dwarf males as sperm delivery systems. They live at depth in the open ocean, mostly from 3,000 to 10,000 feet below the surface, where food is scarce and populations sparse. Females have detached the first dorsal fin ray and moved it forward over their capacious mouth. They dangle a lure at the tip of this spine and literally fish with it. They jiggle and wave the lure while floating, otherwise immobile, in the midst of the sea. The related shallower-water and bottom-dwelling anglerfishes often evolve elaborate mimetic structures for their lures—bits of tissue that resemble worms or even a decoy fish (see essay 3 in *The Panda's Thumb*). Ceratioids live well below the depth that light can penetrate sea water. Their world is one of total ambient darkness, and they must therefore provide the light of attraction themselves. Their lures glow with a luminescence supplied by light glands—a death trap for prey and, perhaps, a beacon for dwarf males.

In 1922, B. Saemundsson, an Icelandic fisheries biologist, dredged a female *Ceratias holbolli,* 26.16 inches in length. To his surprise, he found two small anglerfish, only 2.03 and 2.10 inches long, attached to the female's skin. He assumed, naturally, that they were juveniles, but he was puzzled by their degenerate form: "At first sight," he wrote, "I thought these young ones were pieces of skin torn off and loose." Another oddity puzzled him even more: these small fish were so firmly attached that their lips had grown together about a wad of female tissue projecting well into their mouths and down their throats. Saemundsson could find no other language for his description but an obviously inappropriate mammalian analogy: "The lips are grown together and are attached to a soft papilla or 'teat' protruding,

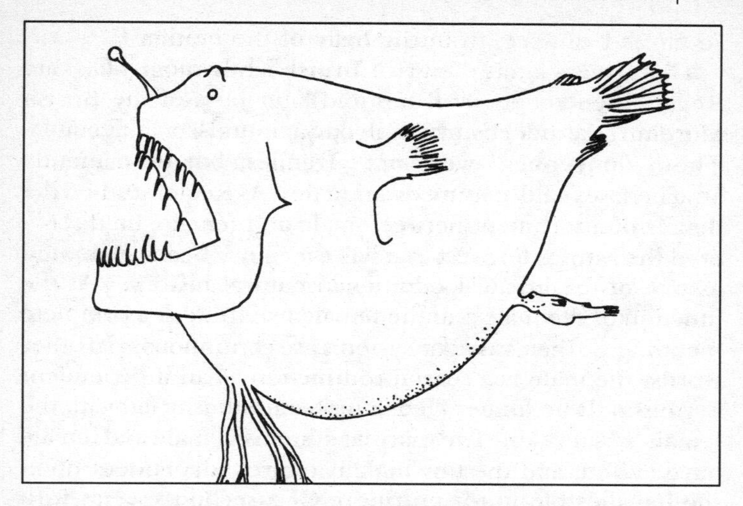

A male anglerfish (lower right), about one-and-a-half inches long, embeds itself into a ten-inch female of the same species. REPRINTED FROM NATURAL HISTORY.

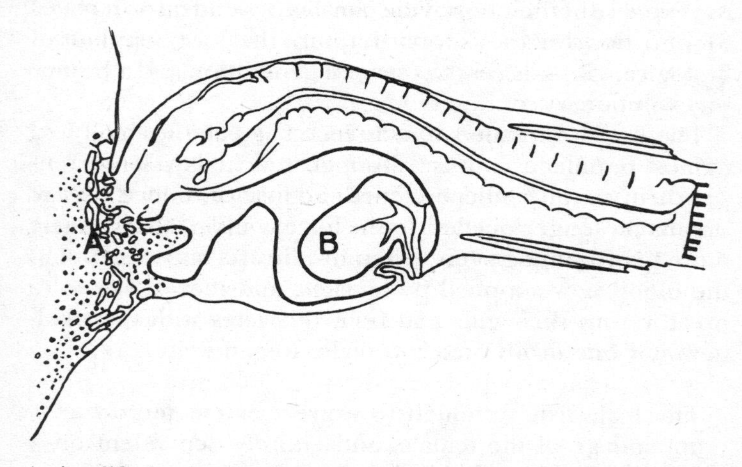

A simplified cross section shows a male anglerfish attached to a female. The two fish share tissue (A), and the male's testis (B), has become enlarged. REPRINTED FROM NATURAL HISTORY.

so far as I can see, from the belly of the mother."

Three years later, the great British ichthyologist C. Tate Regan, then keeper of fishes and later boss of the British Museum (Natural History), solved Saemundsson's dilemma. The "young ones" were not juveniles, but permanently attached, sexually mature dwarf males. As Regan studied the details of attachment between male and female, he discovered the astounding fact that has ever since been celebrated as one of the greatest oddities in natural history: "At the junction of the male and the female fish there is a complete blending . . . their vascular systems are continuous." In other words, the male has ceased to function as an independent organism. It no longer feeds, for its mouth is fused with the female's outer skin. The vascular systems of male and female have united, and the tiny male is entirely dependent upon the female's blood for nutrition. Of a second species with similar habits, Regan writes: "It is impossible to say where one fish begins and the other ends." The male has become a sexual appendage of the female, a kind of incorporated penis. (Both popular and technical literature often refer to the fused male as a "parasite." But I demur. Parasites live at the expense of their host. Fused males depend upon females for nutrition, but they supply in return that most precious of biological gifts—access to the next generation and a chance for evolutionary continuity.)

The extent of male submergence has been exaggerated in most popular accounts. Although attached males surrender their vascular independence and lose or reduce a set of organs no longer needed (eyes, for example), they remain more than a simple penis. Their own hearts must still pump the blood now supplied by females, and they continue to breathe with their gills and remove wastes with their kidneys. Of one firmly attached male, Regan writes:

> The male fish, although to a great extent merely an appendage of the female, and entirely dependent on her for nutrition, yet retains a certain autonomy. He is probably capable, by movements of the tail and fins, of changing his position to some extent. He breathes, he

may have functional kidneys, and he removes from the blood certain products of his own metabolism and keeps them as pigment. . . . But so perfect and complete is the union of husband and wife that one may almost be sure that their genital glands ripen simultaneously, and it is perhaps not too fanciful to think that the female may possibly be able to control the seminal discharge of the male and to ensure that it takes place at the right time for the fertilization of her eggs.

Nonetheless, however autonomous, the males have not honed themselves to Darwinian optimality, for they have evolved no mechanism for excluding other males from subsequent attachment. Several males are often embedded into a single female.

(While criticizing the exaggeration of some popular accounts, allow me a tangential excursion to express a pet peeve. I relied upon primary, technical literature for all my descriptions, but I began by reading several popular renditions. All versions written for nonscientists speak of fused males as the curious tale of *the* anglerfish—just as we so often hear about *the* monkey swinging through trees, or *the* worm burrowing through soil. But if nature teaches any lesson, it loudly proclaims life's diversity. There ain't no such abstraction as *the* clam, *the* fly, or *the* anglerfish. Ceratioid anglerfishes come in nearly 100 species, and each has its own peculiarity. Fused males have not evolved in all species. In some, males attach temporarily, presumably at times of spawning, but never fuse. In others, some males fuse and others become sexually mature while retaining their bodily independence. In still others, fusion is obligatory. In one species of obligate fusers, no sexually mature female has ever been found without an attached male—and the stimulus provided by male hormones may be a prerequisite for maturation.

These obligate fusers have become the paradigm for popular descriptions of *the* anglerfish, but they do not represent the majority of ceratioid species. I grouse because these meaningless abstractions convey seriously false impressions about nature. They greatly exaggerate nature's discontinuities by focusing on extreme forms as false paradigms for an entire group, and rarely mentioning the structurally intermediate species that often live happily and abundantly. If all fishes either had totally independent or completely fused males, then how could we even imagine an evolutionary transition to the peculiar sexual system of *the* anglerfish? But the abundance of structurally intermediate stages—temporary attachment or fusion of some males only—conveys an evolutionary message. These modern structural intermediates are not, of course, actual ancestors of fully fused species, but they do sketch an evolutionary pathway—just as Darwin studied the simple eyes of worms and scallops to learn how a structure so complex and apparently perfect as the vertebrate eye might evolve through a chain of intermediate forms. In any case, bursting diversity is nature's watchword; it should never be submerged by careless abstraction.)

Ceratioid males embark upon their peculiar course early in life. As larvae, they feed normally and live independently. After a period of rapid change, or metamorphosis, males in species destined for fusion do not develop their alimentary canals any further, and never feed again. Their ordinary teeth disappear, and they retain and exaggerate only a few fused teeth at the tips of their mouth—useless in feeding, but well adapted for piercing and holding tight to a female. They become sleek and more streamlined, with a pointed head, compressed body, and strong, propulsive tail fin—in short, a sort of sexual torpedo.

But how do they find females, those tiny dots of connubial matter in the midst of an endless ocean? Most species must use olfactory cues, a system often exquisitely developed in fishes, as in homing salmon that smell out their natal stream. These ceratioid males develop gigantic nostrils after metamorphosis; relative to body size, some ceratioids have larger nasal organs than any other vertebrate. Another family of ceratioids fails to develop large nostrils, but these males have enormously enlarged eyes, and they must search for the ghostly light of fishing females (each species has a different pattern of illumination, and males probably recognize their proper females). The system is not entirely fail-safe, as ichthyologist Ted Pietsch recently found a male of one species attached to a female of a different species—a fatal mistake in evolutionary terms (although the two fish had not fused and might later have separated had not zealous science found and preserved them *in flagrante delicto*).

As I sit here wiggling my toes and flexing my fingers in glorious independence (and with a full one-inch advantage over my wife), I am tempted (but must resist) to apply the standards of my own cherished independence and to pity the poor fused male. It may not be much of a life in our terms, but it keeps several species of anglerfishes going in a strange and difficult environment. And who can judge anyway? In some ultimate Freudian sense, what male could resist the fantasy of life as a penis with a heart, deeply and permanently embedded within a caring and providing female? These anglerfishes represent, in any case, only the extreme expression of nature's more common pattern—smaller males pursuing an evolutionary role as sources of sperm. Do they not, therefore, teach us a generality by their very exaggeration of it? We human males are the oddballs.

I therefore take my leave of fused anglerfishes with a certain sense of awe. Have they not discovered and irrevocably established for themselves what, according to Shakespeare, "every wise man's son doth know"—"journeys end in lovers meeting"?

2 | Nonmoral Nature

WHEN THE Right Honorable and Reverend Francis Henry, earl of Bridgewater, died in February, 1829, he left £8,000 to support a series of books "on the power, wisdom and goodness of God, as manifested in the creation." William Buckland, England's first official academic geologist and later dean of Westminster, was invited to compose one of the nine Bridgewater Treatises. In it he discussed the most pressing problem of natural theology: if God is benevolent and the Creation displays his "power, wisdom and goodness," then why are we surrounded with pain, suffering, and apparently senseless cruelty in the animal world?

Buckland considered the depredation of "carnivorous races" as the primary challenge to an idealized world where the lion might dwell with the lamb. He resolved the issue to his satisfaction by arguing that carnivores actually increase "the aggregate of animal enjoyment" and "diminish that of pain." Death, after all, is swift and relatively painless, victims are spared the ravages of decrepitude and senility, and populations do not outrun their food supply to the greater sorrow of all. God knew what he was doing when he made lions. Buckland concluded in hardly concealed rapture:

> The appointment of death by the agency of carnivora, as the ordinary termination of animal existence, ap-

> pears therefore in its main results to be a dispensation of benevolence; it deducts much from the aggregate amount of the pain of universal death; it abridges, and almost annihilates, throughout the brute creation, the misery of disease, and accidental injuries, and lingering decay; and imposes such salutary restraint upon excessive increase of numbers, that the supply of food maintains perpetually a due ratio to the demand. The result is, that the surface of the land and depths of the waters are ever crowded with myriads of animated beings, the pleasures of whose life are coextensive with its duration; and which throughout the little day of existence that is allotted to them, fulfill with joy the functions for which they were created.

We may find a certain amusing charm in Buckland's vision today, but such arguments did begin to address "the problem of evil" for many of Buckland's contemporaries—how could a benevolent God create such a world of carnage and bloodshed? Yet this argument could not abolish the problem of evil entirely, for nature includes many phenomena far more horrible in our eyes than simple predation. I suspect that nothing evokes greater disgust in most of us than slow destruction of a host by an internal parasite—gradual ingestion, bit by bit, from the inside. In no other way can I explain why *Alien,* an uninspired, grade-C, formula horror film, should have won such a following. That single scene of Mr. Alien, popping forth as a baby parasite from the body of a human host, was both sickening and stunning. Our nineteenth-century forebears maintained similar feelings. The greatest challenge to their concept of a benevolent deity was not simple predation—but slow death by parasitic ingestion. The classic case, treated at length by all great naturalists, invoked the so-called ichneumon fly. Buckland had sidestepped the major issue.

The "ichneumon fly," which provoked such concern among natural theologians, was actually a composite creature representing the habits of an enormous tribe. The Ichneumonoidea are a group of wasps, not flies, that include more species than all the vertebrates combined (wasps, with ants and bees, constitute the order Hymenoptera; flies, with their two wings—wasps have four—form the order Diptera). In addition, many non-ichneumonid wasps of similar habits were often cited for the same grisly details. Thus, the famous story did not merely implicate a single aberrant species (perhaps a perverse leakage from Satan's realm), but hundreds of thousands—a large chunk of what could only be God's creation.

The ichneumons, like most wasps, generally live freely as adults but pass their larval life as parasites feeding on the bodies of other animals, almost invariably members of their own phylum, the Arthropoda. The most common victims are caterpillars (butterfly and moth larvae), but some ichneumons prefer aphids and others attack spiders. Most hosts are parasitized as larvae, but some adults are attacked, and many tiny ichneumons inject their brood directly into the egg of their host.

The free-flying females locate an appropriate host and then convert it to a food factory for their own young. Parasitologists speak of ectoparasitism when the uninvited guest lives on the surface of its host, and endoparasitism when the parasite dwells within. Among endoparasitic ichneumons, adult females pierce the host with their ovipositor and deposit eggs within. (The ovipositor, a thin tube extending backward from the wasp's rear end, may be many times as long as the body itself.) Usually, the host is not otherwise inconvenienced for the moment, at least until the eggs hatch and the ichneumon larvae begin their grim work of interior excavation.

Among ectoparasites, however, many females lay their eggs directly upon the host's body. Since an active host would easily dislodge the egg, the ichneumon mother often simultaneously injects a toxin that paralyzes the caterpillar or other victim. The paralysis may be permanent, and the caterpillar lies, alive but immobile, with the agent of its future destruction secure on its belly. The egg hatches, the helpless caterpillar twitches, the wasp larva pierces and begins its grisly feast.

Since a dead and decaying caterpillar will do the wasp larva no good, it eats in a pattern that cannot help but recall, in our inappropriate, anthropocentric interpretation, the ancient English penalty for treason—drawing and quartering, with its explicit object of extracting as much torment as possible by keeping the victim alive and sentient. As the king's executioner drew out and burned his client's entrails, so does the ichneumon larva eat fat bodies and digestive organs first, keeping the caterpillar alive by preserving intact the essential heart and central nervous system. Finally, the larva completes its work and kills its victim, leaving behind the caterpillar's empty shell. Is it any wonder that ichneumons, not snakes or lions, stood as the paramount challenge to God's benevolence during the heyday of natural theology?

As I read through the nineteenth- and twentieth-century literature on ichneumons, nothing amused me more than the tension between an intellectual knowledge that wasps should not be described in human terms and a literary or emotional inability to avoid the familiar categories of epic and narrative, pain and destruction, victim and vanquisher. We seem to be caught in the mythic structures of our own cultural sagas, quite unable, even in our basic descriptions, to use any other language than the metaphors of battle and conquest. We cannot render this corner of natural history as anything but story, combining the themes of grim horror and fascination and usually ending not so much with pity for the caterpillar as with admiration for the efficiency of the ichneumon.

I detect two basic themes in most epic descriptions: the struggles of prey and the ruthless efficiency of parasites. Although we acknowledge that we may be witnessing little more than automatic instinct or physiological reaction, still we describe the defenses of hosts as though they represented conscious struggles. Thus, aphids kick and caterpillars may wriggle violently as wasps attempt to insert their ovipositors. The pupa of the tortoiseshell butterfly (usually considered an inert creature silently awaiting its conversion from duckling to swan) may contort its abdominal region so sharply that attacking wasps are thrown into the air. The caterpillars of *Hapalia,* when attacked by the wasp *Apanteles machaeralis,* drop suddenly from their leaves and suspend themselves in air by a silken thread. But the wasp may run down the thread and insert its eggs nonetheless. Some hosts can encapsulate the injected egg with blood cells that aggregate and harden, thus suffocating the parasite.

J. H. Fabre, the great nineteenth-century French entomologist, who remains to this day the preeminently literate natural historian of insects, made a special study of parasitic wasps and wrote with an unabashed anthropocentrism about the struggles of paralyzed victims (see his books *Insect Life* and *The Wonders of Instinct*). He describes some imperfectly paralyzed caterpillars that struggle so violently every time a parasite approaches that the wasp larvae must feed with unusual caution. They attach themselves to a silken strand from the roof of their burrow and descend upon a safe and exposed part of the caterpillar:

> The grub is at dinner: head downwards, it is digging into the limp belly of one of the caterpillars. . . . At the least sign of danger in the heap of caterpillars, the larva retreats . . . and climbs back to the ceiling, where the swarming rabble cannot reach it. When peace is restored, it slides down [its silken cord] and returns to table, with its head over the viands and its rear upturned and ready to withdraw in case of need.

In another chapter, he describes the fate of a paralyzed cricket:

> One may see the cricket, bitten to the quick, vainly move its antennae and abdominal styles, open and close its empty jaws, and even move a foot, but the larva is safe and searches its vitals with impunity. What an awful nightmare for the paralyzed cricket!

Fabre even learned to feed paralyzed victims by placing a syrup of sugar and water on their mouthparts—thus showing that they remained alive, sentient, and (by implication) grateful for any palliation of their inevitable fate. If Jesus, immobile and thirsting on the cross, received only vinegar from his tormentors, Fabre at least could make an ending bittersweet.

The second theme, ruthless efficiency of the parasites, leads to the opposite conclusion—grudging admiration for the victors. We learn of their skill in capturing dangerous hosts often many times larger than themselves. Caterpillars may be easy game, but psammocharid wasps prefer spiders. They must insert their ovipositors in a safe and precise spot. Some leave a paralyzed spider in its own burrow. *Planiceps hirsutus,* for example, parasitizes a California trapdoor spider. It searches for spider tubes on sand dunes, then digs into nearby sand to disturb the spider's home and drive it out. When the spider emerges, the wasp attacks, paralyzes its victim, drags it back into its own tube, shuts and fastens the trapdoor, and deposits a single egg upon the spider's abdomen. Other psammocharids will drag a heavy spider back to a previously prepared cluster of clay or mud cells. Some amputate a spider's legs to make the passage easier. Others fly back over water, skimming a buoyant spider along the surface.

Some wasps must battle with other parasites over a host's body. *Rhyssella curvipes* can detect the larvae of wood wasps deep within alder wood and drill down to a potential victim with its sharply ridged ovipositor. *Pseudorhyssa alpestris,* a related parasite, cannot drill directly into wood since its slender ovipositor bears only rudimentary cutting ridges. It locates the holes made by *Rhyssella,* inserts its ovipositor, and lays an egg on the host (already conveniently paralyzed by *Rhyssella*), right next to the egg deposited by its relative. The two eggs hatch at about the same time, but the larva of *Pseudorhyssa* has a bigger head bearing much larger mandibles. *Pseudorhyssa* seizes the smaller *Rhyssella* larva, destroys it, and proceeds to feast upon a banquet already well prepared.

Other praises for the efficiency of mothers invoke the themes of early, quick, and often. Many ichneumons don't even wait for their hosts to develop into larvae, but parasitize the egg directly (larval wasps may then either drain the egg itself or enter the developing host larva). Others simply move fast. *Apanteles militaris* can deposit up to seventy-two eggs in a single second. Still others are doggedly persistent. *Aphidius gomezi* females produce up to 1,500 eggs and can parasitize as many as 600 aphids in a single working day. In a bizarre twist upon "often," some wasps indulge in polyembryony, a kind of iterated supertwining. A single egg divides into cells that aggregate into as many as 500 individuals. Since some polyembryonic wasps parasitize caterpillars much larger than themselves and may lay up to six eggs in each, as many as 3,000 larvae may develop within, and feed upon a single host. These wasps are endoparasites and do not paralyze their victims. The caterpillars writhe back and forth, not (one suspects) from pain, but merely in response to the commotion induced by thousands of wasp larvae feeding within.

Maternal efficiency is often matched by larval aptitude. I have already mentioned the pattern of eating less essential parts first, thus keeping the host alive and fresh to its final and merciful dispatch. After the larva digests every edible morsel of its victim (if only to prevent later fouling of its abode by decaying tissue), it may still use the outer shell of its host. One aphid parasite cuts a hole in the bottom of its victim's shell, glues the skeleton to a leaf by sticky secretions from its salivary gland, and then spins a cocoon to pupate within the aphid's shell.

In using inappropriate anthropocentric language for this romp through the natural history of ichneumons, I have tried to emphasize just why these wasps became a preeminent challenge to natural theology—the antiquated doctrine that attempted to infer God's essence from the products of his creation. I have used twentieth-century examples for the most part, but all themes were known and stressed by the great nineteenth-century natural theologians. How then did they square the habits of these wasps with the goodness of God? How did they extract themselves from this dilemma of their own making?

The strategies were as varied as the practitioners; they shared only the theme of special pleading for an a priori doctrine—our naturalists *knew* that God's benevolence was lurking somewhere behind all these tales of apparent horror. Charles Lyell, for example, in the first edition of his epochal *Principles of Geology* (1830–1833), decided that caterpillars posed such a threat to vegetation that any natural checks upon them could only reflect well upon a creating deity, for caterpillars would destroy human agriculture "did not Providence put causes in operation to keep them in due bounds."

The Reverend William Kirby, rector of Barham, and Britain's foremost entomologist, chose to ignore the plight of caterpillars and focused instead upon the virtue of mother love displayed by wasps in provisioning their young with such care.

> The great object of the female is to discover a proper nidus for her eggs. In search of this she is in constant motion. Is the caterpillar of a butterfly or moth the appropriate food for her young? You see her alight upon the plants where they are most usually to be met with, run quickly over them, carefully examining every leaf, and, having found the unfortunate object of her search, insert her sting into its flesh, and there deposit an egg. . . . The active Ichneumon braves every danger, and does not desist until her courage and address have insured subsistence for one of her future progeny.

Kirby found this solicitude all the more remarkable because the female wasp will never see her child and enjoy the pleasures of parenthood. Yet love compels her to danger nonetheless:

> A very large proportion of them are doomed to die before their young come into existence. But in these the passion is not extinguished. . . . When you witness

> the solicitude with which they provide for the security and sustenance of their future young, you can scarcely deny to them love for a progeny they are never destined to behold.

Kirby also put in a good word for the marauding larvae, praising them for their forbearance in eating selectively to keep their caterpillar alive. Would we all husband our resources with such care!

> In this strange and apparently cruel operation one circumstance is truly remarkable. The larva of the Ichneumon, though every day, perhaps for months, it gnaws the inside of the caterpillar, and though at last it has devoured almost every part of it except the skin and intestines, carefully all this time it avoids injuring the vital organs, as if aware that its own existence depends on that of the insect upon which it preys! . . . What would be the impression which a similar instance amongst the race of quadrupeds would make upon us? If, for example, an animal . . . should be found to feed upon the inside of a dog, devouring only those parts not essential to life, while it cautiously left uninjured the heart, arteries, lungs, and intestines,—should we not regard such an instance as a perfect prodigy, as an example of instinctive forbearance almost miraculous? [The last three quotes come from the 1856, and last pre-Darwinian, edition of Kirby and Spence's *Introduction to Entomology.*]

This tradition of attempting to read moral meaning from nature did not cease with the triumph of evolutionary theory in 1859—for evolution could be read as God's chosen method of peopling our planet, and ethical messages might still populate nature. Thus, St. George Mivart, one of Darwin's most effective evolutionary critics and a devout Catholic, argued that "many amiable and excellent people" had been misled by the apparent suffering of animals for two reasons. First, whatever the pain, "physical suffering and moral evil are simply incommensurable." Since beasts are not moral agents, their feelings cannot bear any ethical message. But secondly, lest our visceral sensitivities still be aroused, Mivart assures us that animals must feel little, if any, pain. Using a favorite racist argument of the time—that "primitive" people suffer far less than advanced and cultured folk—Mivart extrapolated further down the ladder of life into a realm of very limited pain indeed: Physical suffering, he argued,

> depends greatly upon the mental condition of the sufferer. Only during consciousness does it exist, and only in the most highly organized men does it reach its acme. The author has been assured that lower races of men appear less keenly sensitive to physical suffering than do more cultivated and refined human beings. Thus only in man can there really be any intense degree of suffering, because only in him is there that intellectual recollection of past moments and that anticipation of future ones, which constitute in great part the bitterness of suffering. The momentary pang, the present pain, which beasts endure, though real enough, is yet, doubtless, not to be compared as to its intensity with the suffering which is produced in man through his high prerogative of self-consciousness [from *Genesis of Species,* 1871].

It took Darwin himself to derail this ancient tradition—and he proceeded in the gentle way so characteristic of his radical intellectual approach to nearly everything. The ichneumons also troubled Darwin greatly and he wrote of them to Asa Gray in 1860:

> I own that I cannot see as plainly as others do, and as I should wish to do, evidence of design and beneficence on all sides of us. There seems to me too much misery in the world. I cannot persuade myself that a beneficent

> and omnipotent God would have designedly created the Ichneumonidae with the express intention of their feeding within the living bodies of Caterpillars, or that a cat should play with mice.

Indeed, he had written with more passion to Joseph Hooker in 1856: "What a book a devil's chaplain might write on the clumsy, wasteful, blundering, low, and horribly cruel works of nature!"

This honest admission—that nature is often (by our standards) cruel and that all previous attempts to find a lurking goodness behind everything represent just so much special pleading—can lead in two directions. One might retain the principle that nature holds moral messages, but reverse the usual perspective and claim that morality consists in understanding the ways of nature and doing the opposite. Thomas Henry Huxley advanced this argument in his famous essay on *Evolution and Ethics* (1893):

> The practice of that which is ethically best—what we call goodness or virtue—involves a course of conduct which, in all respects, is opposed to that which leads to success in the cosmic struggle for existence. In place of ruthless self-assertion it demands self-restraint; in place of thrusting aside, or treading down, all competitors, it requires that the individual shall not merely respect, but shall help his fellows. . . . It repudiates the gladiatorial theory of existence. . . . Laws and moral precepts are directed to the end of curbing the cosmic process.

The other argument, radical in Darwin's day but more familiar now, holds that nature simply is as we find it. Our failure to discern a universal good does not record any lack of insight or ingenuity, but merely demonstrates that nature contains no moral messages framed in human terms. Morality is a subject for philosophers, theologians, students of the humanities, indeed for all thinking people. The answers will not be read passively from nature; they do not, and cannot, arise from the data of science. The factual state of the world does not teach us how we, with our powers for good and evil, should alter or preserve it in the most ethical manner.

Darwin himself tended toward this view, although he could not, as a man of his time, thoroughly abandon the idea that laws of nature might reflect some higher purpose. He clearly recognized that specific manifestations of those laws—cats playing with mice, and ichneumon larvae eating caterpillars—could not embody ethical messages, but he somehow hoped that unknown higher laws might exist "with the details, whether good or bad, left to the working out of what we may call chance."

Since ichneumons are a detail, and since natural selection is a law regulating details, the answer to the ancient dilemma of why such cruelty (in our terms) exists in nature can only be that there isn't any answer—and that framing the question "in our terms" is thoroughly inappropriate in a natural world neither made for us nor ruled by us. It just plain happens. It is a strategy that works for ichneumons and that natural selection has programmed into their behavioral repertoire. Caterpillars are not suffering to teach us something; they have simply been outmaneuvered, for now, in the evolutionary game. Perhaps they will evolve a set of adequate defenses sometime in the future, thus sealing the fate of ichneumons. And perhaps, indeed probably, they will not.

Another Huxley, Thomas's grandson Julian, spoke for this position, using as an example—yes, you guessed it—the ubiquitous ichneumons:

> Natural selection, in fact, though like the mills of God in grinding slowly and grinding small, has few other attributes that a civilized religion would call divine. . . . Its products are just as likely to be aesthetically, morally, or intellectually repulsive to us as they are to be attractive. We need only think of the ugliness of *Sacculina* or a bladder-worm, the stupidity of a rhinoceros or a stego-

saur, the horror of a female mantis devouring its mate or a brood of ichneumon flies slowly eating out a caterpillar.

If nature is nonmoral, then evolution cannot teach any ethical theory at all. The assumption that it can has abetted a panoply of social evils that ideologues falsely read into nature from their beliefs—eugenics and (misnamed) social Darwinism prominently among them. Not only did Darwin eschew any attempt to discover an antireligious ethic in nature, he also expressly stated his personal bewilderment about such deep issues as the problem of evil. Just a few sentences after invoking the ichneumons, and in words that express both the modesty of this splendid man and the compatibility, through lack of contact, between science and true religion, Darwin wrote to Asa Gray,

> I feel most deeply that the whole subject is too profound for the human intellect. A dog might as well speculate on the mind of Newton. Let each man hope and believe what he can.

Postscript

Michele Aldrich sent an even better literary reference than any I had found. Mark Twain, in a biting bit of satire called "Little Bessie Would Assist Providence," chronicles a conversation of mother and daughter—daughter insisting that a benevolent God would not have given her little friend "Billy Norris the typhus" and visited other unjust disasters upon decent people, mother assuring her that there must be a good reason for it all. Bessie's last rejoinder, which summarily ends the essay as you shall see, invokes our old friends, the ichneumons:

> Mr. Hollister says the wasps catch spiders and cram them down into their nests in the ground—alive, mama!—and there they live and suffer days and days

and days, and the hungry little wasps chewing their legs and gnawing into their bellies all the time, to make them good and religious and praise God for His infinite mercies. I think Mr. Hollister is just lovely, and ever so kind; for when I asked him if he would treat a spider like that he said he hoped to be damned if he would; and then he—Dear mama, have you fainted!

James W. Tuttleton, chairman of the English department at New York University, sent me a stunning poem by Robert Frost that seems designed as a commentary upon Darwin's last statement that chance may regulate in the small, even if purpose might be found in the large. Or do we even see true purpose in the large? The poem is called, simply, "Design":

I found a dimpled spider, fat and white,
On a white heal-all, holding up a moth
Like a white piece of rigid satin cloth—
Assorted characters of death and blight
Mixed ready to begin the morning right,
Like the ingredients of a witches' broth—
A snow-drop spider, a flower like a froth,
And dead wings carried like a paper kite.

What had that flower to do with being white,
The wayside blue and innocent heal-all?
What brought the kindred spider to that height,
Then steered the white moth thither in the night?
What but design of darkness to appall?—
If design govern in a thing so small.

I was so struck by the image of the spider as a drop, the flower as a froth, the moth as a pair of two-dimensional wings. Forms so unlike, yet all white and all brought together in one spot for destruction. Why? Or, as we read the last two lines, may we even ask such a question? I think that we cannot, and I regard this insight as the most liberating theme of Darwin's revolution.

3 | The Guano Ring

WHEN I FIRST WENT to sea as a petrified urbanite who had never ridden anything larger than a rowboat, an old sailor (and Navy man) told me that I could chart my way through this *aqua incognita* if I remembered but one simple rule for life and work aboard a ship: if it moves, salute it; if it doesn't move, paint it.

If we analyze why such a statement counts as a joke (albeit a feeble one) in our culture, we must cite the incongruity of placing such a "mindless" model for making decisions inside a human skull. After all, the essence of human intelligence is creative flexibility, our skill in grasping new and complex contexts—in short, our ability to make (as we call them) judgments, rather than to act by the dictates of rigid, preset rules. We are, as Konrad Lorenz has stated, "specialists in nonspecialization." We do not behave as machines with simple yes-no switches, invariably triggered by definite bits of information present in our immediate environments. Our enlightened sailor, no matter how successful at combating rust or avoiding the brig, is not following a human style of intelligence.

Yet this inflexible model does represent the style of intelligence followed with great success by most other animals. The decisions of animals are usually unambiguous yeses or noes triggered by definite signals, not subtle choices based upon the assessment of a complex gestalt.

Many birds, for example, do not recognize their own

young and act instead by the rule: care for what is inside the nest; ignore what is outside. British ethologist W. H. Thorpe writes: "Most birds, while they may be very attentive to their young in the nest, are completely callous and unresponsive to those same young when, as a result of some accident, they are outside the nest or the immediate nest territory."

This rule rarely poses evolutionary dilemmas for birds, since the objects in their nests are usually their own young (carrying their Darwinian heritage of shared genes). But this inflexible style of intelligence can be exploited and commandeered to a nefarious purpose by other species. Cuckoos, for example, lay their eggs in the nests of other birds. A cuckoo hatchling, usually larger and more vigorous than the rightful inhabitants, often expels its legitimate nest mates, which then die, frantically begging for food, while their parents follow the rule: ignore them for their inappropriate location, and feed the young cuckoo instead. We can intellectualize our anthropomorphism away, but we cannot expunge it from our aesthetic reactions. I must confess that no scene of organic activity makes me angrier about the world's injustice than the sight of a foster parent, its own young killed by a cuckoo, solicitously feeding a begging parasite that may be up to five times its own size (cuckoos often choose much smaller birds as their hosts, and the fledglings may be much larger than their foster parents).

During a recent trip to the Galápagos Islands, I encountered another, interestingly different, example of birds that twist this common rule to different uses. This time, both the victim and benefactor are true siblings and the end result, although condemning weaker siblings to death, is evolutionary advantage for family lines.

The boobies (along with their cousins, the gannets) form a small (nine species) but widespread family of seabirds, the Sulidae. (Everything, and more, that you will ever want to know about sulids you will find in J. Bryan Nelson's magnificent monograph: *The Sulidae: Gannets and Boobies,* 1978.) Earliest references in the Oxford English Dictionary indicate that boobies received their unflattering name, not for

A blue-footed booby incubates an egg inside the guano ring marking its "nest." Galápagos, North Seymour Island. PHOTO BY DUNCAN M. PORTER.

the distinctive waddling walk of one major display, big feet out and head held high in a behavior called "sky pointing," but for their remarkable tameness, which allowed sailors (bent only on destruction) to catch them so easily.

Three species of sulids inhabit the Galápagos Islands: the red-footed, the blue-footed, and the masked booby. The red-footed booby lays a single egg in a conventional nest built near the tops and edges of trees and bushes. By contrast, its cousin of markedly different natural pedicure, the blue-footed booby, lays its eggs on the ground and builds no true nest at all. Instead, it delimits the nesting area in a remarkable and efficient way: it squirts guano (birdshit to nonornithologists who have not read *Doctor No*) in all directions around itself, thus producing a symmetrical white ring as a symbolic marker of its nest.

Within this ring, the female blue-foot lays, not one (as in many boobies), but from one to three eggs. In his most impressive discovery, Nelson has explained much about the breeding behavior and general ecology of boobies by linking the production of eggs and young to the quality and style of feeding in parents. Boobies that travel long distances (up to 300 miles) to locate scarce sources of food, tend to lay but a single large egg, hatching into a resilient chick that can survive long intervals between parental feedings. On the other hand, when food sources are rich, dependable, and near, more eggs are laid and more young reared. At the extreme of this tendency lies the Peruvian booby, with its clutch of two to four eggs (averaging three) and its ability to raise all chicks to adulthood. Peruvian boobies feed on the teeming anchovies of their local waters, fish that may be almost as densely packed in the ocean as in the sardine cans that may become their posthumous home.

The blue-foot lies between these two tendencies. It is a nearshore feeder, but its sources have neither the richness nor the predictability of swarming anchovies. Consequently, conditions vary drastically from generation to generation. The blue-foot has therefore evolved a flexible strategy based on the exploitation by older siblings of their parents' intellectual style: yes-no decisions triggered by simple signals. In good times, parents may lay up to three eggs and successfully fledge all three chicks; in poor years, they may still lay two or three eggs and hatch all their chicks, but only one can survive. The death of nest (or, rather, ring) mates is not the haphazard result of a losing struggle to feed all chicks with insufficient food, but a highly systematic affair based on indirect murder by the oldest sibling.

I was reminded of the quip about painting and saluting while observing blue-footed boobies on Hood Island in the Galápagos. Their guano rings cover the volcanic surface in many places, often blocking the narrow paths that visitors must tread in these well protected islands. Parents sit on their eggs and young chicks, apparently oblivious to groups of visitors who gawk, gesticulate, and point cameras within inches of their territory. Yet I noticed, at first by accident, that any intrusion into a guano ring would alter the behavior of adult birds from blissful ignorance to directed aggression. A single toe across the ring elicited an immediate barrage of squawking, posturing, and pecking. A few casual experiments led me to the tentative conclusion that the boundary is an invisible circle right in the middle of the ring. I could cautiously advance my toe across the outer part of the ring with no effect; but as I moved it forward, as slowly and as unobtrusively as possible, I invariably passed a central point that brought on the pronounced parental reaction all at once.

Three hours later, I learned from our excellent guides and from Bryan Nelson's popular book (*Galápagos: Islands of Birds*), how older siblings exploit this parental behavior. And, anthropomorphic as we all must be, it sent a shiver of wonder and disgust up my spine. (Science, to a large extent, consists of enhancing the first reaction and suppressing the second.) The female blue-foot lays her eggs several days apart, and they hatch in the same order. The firstborn sibling is thus larger and considerably stronger than its one or two ring mates. When food is abundant, parents feed all chicks adequately and the firstborn does not molest its younger siblings. But when food is scarce and only one chick can survive, the actions of younger sibs evoke (how, we do not know) a different behavior by big sister or brother. The oldest simply pushes its younger siblings outside the guano ring. As human mammals, our first reaction might be: so what? The younger sibs are not physically hurt and they end up but a few inches from the ring, where parents will surely notice their plaintive sounds and struggling motions and gather them quickly back.

But a parental booby does no such thing, for it operates like our proverbial sailor who made an either-or judgment by invoking the single criterion of movement. Parental boobies work by the rule: if a chick is inside the ring, care for it; if it is outside, ignore it. Even if the chick should flop, by happenstance, upon the ring, it will be rejected with all the vehemence applied to my transgressing toe.

We saw a chick on Hood Island struggling just a foot outside the ring in plain sight of the parent within, sitting (in an attitude that we tend to read as maternal affection) upon the triumphant older sibling (which did not, however, seem to be smirking). Every mother's son and daughter among us longed to replace the small chick, but a belief in noninterference must be respected even when it hurts. For if we understand this system aright, such a slaughter of the innocents is a hecatomb for success of the lineages practicing it. Older chicks only expel their siblings when food cannot be secured to raise them all. A parental struggle to raise three on food for one would probably lead to the death of all.

The rule of "nurture within, ignore or reject outside" cannot represent all the complexity of social behavior in nesting boobies. After all, most birds are noted "egalitarians" in their division of labor between sexes, and male boobies are almost as attentive as females in incubating eggs and chicks. Since each brooding stint lasts about a day, boobies must permit their mates to transgress the sanctity of the guano ring when exchanging roles of care and provision. Still, the basic rule remains in force; it is not flouted but rather overridden by specific and recognized signals that act as a ticket of admission. K. E. L. Simmons, working on Ascension Island with the related brown booby, described the extensive series of calls and landing rituals that returning mates use to gain admission to their territory. But when adults trespass upon the unattended territory of an unrelated bird (as they often do to scrounge nest material on the cheap), they enter as "silently and as inconspicuously as possible."

If chicks could perform the overriding behavior, they too could win readmission to the ring. Indeed, they learn these signals as they age, as well they must, for older chicks begin to wander from the ring as they gain sufficient mobility for such travels at about four to five weeks of age. (Nelson argues that they wander primairly to seek shade when both parents are foraging; overheating is a primary cause of death in booby chicks.) Yet hatchling boobies display only a few behaviors—little more than food begging and bill hiding (appeasement) gestures, as Nelson demonstrates—and the overriding signals for entrance into the ring are not among them.

The third species of the Galápagos, the white, or masked, booby, works on a more rigid system, but follows the same rules as its blue-footed cousin. Masked boobies are distant foragers, feeding primarily on flying fish. By Nelson's maxim, they should be able to raise but one chick. Sometimes, masked boobies lay only one egg, but usually they provision each nesting site with two. In this case, "brood reduction" (to use the somewhat euphemistic jargon) is obligatory. The older chick always pushes its younger sibling outside the nest (or occasionally stomps it to death within). This system seems, at first, to make no sense. The blue-foots, whatever our negative, if inappropriate, emotional reactions, at least use sibling murder as a device to match the number of chicks to a fluctuating supply of available food.

By what perverse logic should masked boobies produce two eggs, yet never rear more than one chick invariably branded with the mark of Cain? Nelson argues forcefully that clutches of two eggs represent an adaptation for greatly increased success in raising *one* chick. The causes of death in eggs and young hatchlings are numerous—siblings intent upon murder being only one of many dangers to which booby flesh is heir. Eggs crack or roll from the nest; tiny hatchlings easily overheat. The second egg may represent insurance against death of the first chick. A healthy first chick cancels the policy directly, but the added investment may benefit parents as a hedge worth the expense of producing another egg (they will, after all, never need to expend much energy in feeding an unnecessary second chick). At Kure Atoll in the Hawaiian Archipelago, for example, clutches of two eggs successfully fledged one chick in 68 percent of nests examined during three years. But clutches of one egg fledged their single chick only 32 percent of the time.

Evolutionary biologists, by long training and ingrained

habit, tend to discuss such phenomena as the siblicide of boobies in the language of adaptation: how does a behavior that seems, at first sight, harmful and irrational really represent an adaptation finely honed by natural selection for the benefit of struggling individuals? Indeed, I have (and somewhat uncharacteristically for me) used the conventional language in this essay, for Nelson's work persuades me that siblicide is a Darwinian adaptation for maximizing the success of parents in rearing the largest number of chicks permitted by prevailing abundances of food.

But I am most uncomfortable in attributing the basic behavioral style, which permits siblicide as a specific manifestation, only to adaptation, although this too is usually done. I speak here of the basic mode of intelligence that permits siblicide to work: the sailor's system (of my opening paragraph) based on yes-no decisions triggered by definite signals. John Alcock, for example, in a leading contemporary text (*Animal Behavior: An Evolutionary Approach,* 1975) argues over and over again that this common intellectual style is, in itself and in general, an adaptation directly fashioned by natural selection for optimal responses in prevailing environments: "Programmed responses are widespread," he writes, "because animals that base their behavior on relatively simple signals provided by important objects in their environment are likely to do the biologically proper thing."

(On the overwhelming power of natural selection, no less a personage than H.R.H. Prince Philip, duke of Edinburgh, has written in the preface to Nelson's popular book on birds of the Galápagos: "The process of natural selection has controlled the very minutest detail of every feature of the whole individual and the group to which it belongs." I do not cite this passage facetiously to win an argument by saddling a position I do not accept with a mock seal of royal approval, but rather to indicate how widely the language of strict adaptation has moved beyond professional circles into the writing of well-informed amateurs.)

As I argued for siblicide and guano rings, I am prepared to view any specific manifestation of my sailor's intellectual style as an adaptation. But I cannot, as Alcock claims, view the style itself as no more than the optimized product of unconstrained natural selection. The smaller brain and more limited neural circuitry of nonhuman animals must impose, or at least encourage, intellectual modes different from our own. These smaller brains need not be viewed as direct adaptations to any prevailing condition. They represent, rather, inherited structural constraints that limit the range of specific adaptations fashioned within their orbit. The sailor's style is a constraint that permits boobies to reduce their broods by exploiting a behavioral repertoire based on inflexible rules and simple triggers. Such a system would not work in humans, for parents will not cease to recognize their babies after a small and simple change in location. In human societies that practice infanticide (for ecological reasons often quite similar to those inducing siblicide in boobies), explicit social rules or venerated religious traditions—rather than mere duplicity by removal—must force or persuade parental action.

Birds may have originally developed their brain, with its characteristic size, as an adaptation to life in an ancestral lineage more than 200 million years ago; the sailor's style of intelligence may be a nonadaptive consequence of this inherited design. Yet this style has set the boundaries of behavior ever since. Each individual behavior may be a lovely adaptation, but it must be fashioned within a prevailing constraint. Which is more important: the beauty of the adaptation or the constraint that limits it to a permissible path? We cannot and need not choose, for both factors define an essential tension that regulates all evolution.

The sources of organic form and behavior are manifold and include at least three primary categories. We have just discussed two: immediate adaptations fashioned by natural selection (exploitation by older booby siblings of their parents' intellectual style, leading to easy dispatch of nest mates); and potentially nonadaptive consequences of basic structural designs acting as constraints upon the pathways of adaptation (the intellectual style of yes-no decisions based on simple triggers).

In a third category, we find definite ancestral adaptations now used by descendants in different ways. Nelson has shown, for example, that boobies reinforce the pair-bond between male and female through a complex series of highly ritualized behaviors that include gathering objects and presenting them to mates. In boobies that lay their eggs upon the ground, these behaviors are clearly relics of actions that once served to gather material for ancestral nests—for some of the detailed motions that still build nests in related species are followed, while others have been lost. The egg-laying areas of masked boobies are strewn with appropriate bits of twigs and other nesting materials that adults gather for their mutual displays and then must sweep out of the guano ring to lie unused upon the ground. I have emphasized these curious changes in function in several other essays (see 4 and 11) for they are the primary proof of evolution—forms and actions that only make sense in the light of a previous, inherited history.

When I wonder how three such disparate sources can lead to the harmonious structures that organisms embody, I temper my amazement by remembering the history of languages. Consider the amalgam that English represents—vestiges, borrowings, fusions. Yet poets continue to create things of beauty. Historical pathways and current uses are different aspects of a common subject. The pathways are intricate beyond all imagining, but only the hearty travelers remain with us.

4 Quick Lives and Quirky Changes

POSTHUMOUS TRIUMPH IS HOLLOW, however abstractly rewarding. Nanki-Poo refused Ko-Ko's inducement to undergo a ceremonious public beheading rather than a private suicide: "There'll be a procession—bands—dead march—bells tolling . . . then, when it's all over, general rejoicings, and a display of fireworks in the evening. You won't see them, but they'll be there all the same." And I never could figure out why America's premier nineteenth-century anthropologists J. W. Powell and WJ McGee made a bet about who had the larger brain—to be settled by autopsy when the joy of victory could no longer be savored.

Nonetheless, I just made a dumb bet with a female jogging enthusiast: that no woman would win the Boston marathon in my lifetime. I'd rather lose, but expect I won't. Still, if superior average speed of running males is among the few insignificant but genuinely biological differences between human sexes, I can only respond to charges of gloating (for the abstraction I represent, but not, alas, for me and my huffing eight-minute miles) with a statement of genuine regret; how gladly would I trade this useless advantage for the most precious benefit of being female—several extra years of average life.

I do not know whether shorter male life is a generality in nature—and whether we should therefore add to smaller average size (see essay 1) another biological strike against machismo—but I just learned (with thanks to Martin L. Adamson) about an instructive extreme case.

In 1962, James H. Oliver Jr. traced the life cycle of a mite that parasitizes the cocoons of earthworms. Both males and females of *Histiostoma murchiei* pass through an egg and three juvenile stages before molting into an adult. In addition, the female intercalates one additional stage—euphoniously named the hypopus—between the second and third pre-adult phases. Females develop at a leisurely pace for such a small creature. Discounting the hypopus, the passage from egg to adult, through stages held in common with males, takes one to three weeks. The additional hypopus may extend female life greatly—for these mites find and infest other cocoons only during the hypopal stage (males always stay at home). The hypopus may, first of all, remain dormant for long periods within the skin of the previous juvenile stage, awaiting (so to speak) favorable conditions for emergence and movement to another cocoon. When the hypopus does emerge, it may then live for a long time, moving about in its own cocoon (and sometimes becoming dormant again) or moving out in search of a new home.

Males, by contrast, race through the same stages (minus the hypopus) with a celerity that should inspire Bill Rodgers as he trudges up Heartbreak Hill next Patriot's Day. "Adult males," Oliver writes, "have been observed copulating with their mother within 3 to 4 days after being laid as eggs"—and they die soon after this bout of incestuous joy. Why this outstanding difference in life-span between the sexes? And what has it to do with the Oedipal habits of these mites? A further look at the unusual reproductive biology of these parasites seems to provide the answer.

When a hypopus finds a new cocoon, it lays two to nine eggs within two days after molting into an adult—and without benefit of fertilization. All these eggs develop into males—the only source of potential husbands as well. What better evolutionary rationale for rapid male development could we hope to find? The females of most species must seek their husbands. These mites make them from scratch and then wait. Males of *Histiostoma murchiei* are little more than sources of sperm; the sooner they can perform, the better.

Two days after its incestuous mating, the female begins to lay eggs again and may continue for two to five days, producing as many as 500 offspring—all female this time.

In solving one problem—the differential speed of development between sexes—we have only encountered a more curious question: how can this system work in the first place; how can an unmated female, alone in a new cocoon, produce a generation of husbands, and why are the offspring of her next reproductive bout all female?

The answer to this broader question lies in the unfamiliar style of sex determination in these mites. In most animals, both males and females have paired chromosomes, and the status of one pair determines the sex of its bearer. Human females, for example, have two large sex chromosomes (designated XX), while males have one large (X) and one small (Y) chromosome in their determining pair. All unfertilized egg cells carry a single X, while sperm carry either an X or a Y. We each owe our sex to the good fortune of one sperm among the millions per ejaculate. Animals with paired chromosomes in both sexes are called diploid.

Some animals use a different system of sex determination. Females are diploid, but males have only one chromosome for each female pair and are called haploid (for half the diploid number). Males, in other words—and ironic as this may seem—develop from unfertilized eggs and have no fathers. Fertilized eggs produce diploid females. Animals using this system are called haplodiploid (because males are haploid and females diploid).

Histiostoma murchiei is haplodiploid. Hence, the unmated female in a new cocoon raises a generation of males from unfertilized eggs, and a subsequent generation of females from the resulting incest.

Haplodiploidy, a fascinating phenomenon rich in implication, has circulated through these essays in various contexts for years. It helped to explain the origin of social systems in ants and bees (see essay 33 in *Ever Since Darwin*), and it underlay the habits of a male mite who fertilizes several sisters within his mother's body, and dies before "birth" (essay 6 in *The Panda's Thumb*). It also circulates widely through the animal kingdom. Haplodiploid species have been found in rotifers, nematodes, mites, and in four separate orders of insects—the Thysanoptera (thrips), the Homoptera (aphids, cicadas, and their allies), the Coleoptera (beetles), and the Hymenoptera (ants, bees, and wasps). These groups are not closely related and their presumed common ancestors are diploid. Thus, haplodiploidy has arisen independently—and often many times—within each group. Although most of these groups contain only a few haplodiploid species amidst a host of ordinary diploids, the Hymenoptera, with more than 100,000 named species, are exclusively haplodiploid. Since vertebrates only include some 50,000 species, as Oliver reminds us, our chauvinistic impression that haplodiploidy is curious or rare should also be revised. At least 10 percent of all named animal species are haplodiploid.

Within the last decade, haplodiploidy has figured most prominently in the news (both general and scientific) for its role in an ingenious Darwinian explanation of an old biological mystery—the origin of sociality in Hymenoptera, particularly the existence of sterile "worker" castes, invariably female, in ants and bees. Since sociality evolved several times within the Hymenoptera, the invariant system of sterile female castes demands a general explanation. The larger problem is even more puzzling: why, in a presumably Darwinian world filled with organisms acting only for their personal reproductive success, should large numbers of females "forego" their own reproduction to help their mother (the queen) raise more sisters?

The ingenious explanation relies upon the peculiar asymmetries of genetic relationship between sexes in haplodiploid animals. In both diploids and haplodiploids, mothers pass half their genetic material (one set of chromosomes in each egg cell) to each offspring. They are therefore equally related (by half of their genetic selves) to both sons and daughters. A female in diploid species also shares approximately half her genes with both brothers and sisters. But a female in haplodiploid species shares three-quarters of her genes with sisters and only one-quarter with brothers for the following reason: Consider any gene (one copy on a single chromosome) in sisters. What is the probability that a brother will share it? If the gene is on a paternal chromosome, then the brother has zero probability of sharing it, for he has no paternal chromosomes. If the gene is on a maternal chromosome, then he has a 50 percent chance of sharing it with his sister—because he either received the same chromosome from his mother, or the other member of the pair. Thus, summing over all genes, the relationship between brother and sister is the average between zero (for paternal genes of sisters, necessarily absent in brothers) and 50 percent (for maternal genes)—or 25 percent.

What then is the probability that a sister will share the same gene? If it is a paternal gene, the sister must share it since fathers have only one set of chromosomes and they pass their entire genetic program to each daughter. If it is a maternal gene, the chance is 50 percent by the same argument advanced for brothers. The total relationship between sisters is therefore the average between 100 percent (for paternal genes) and 50 percent (for maternal genes)—or 75 percent.

Females are therefore more closely related to their sisters (by three-quarters) than either to their mothers (by one-half) or to their own potential offspring (also by one-half). If the Darwinian imperative leads organisms to maximize the numbers of their own genes in future generations, then females will do better by helping their mother raise sisters (as sterile workers do) than by producing their own offspring. Thus, the asymmetry of genetic relationship in haplodiploids may explain both why worker castes of social Hymenoptera are invariably female, and why sociality in this style has evolved many times among the Hymenoptera, but not among the much larger array of diploid organisms. (As always, our complex world provides an exception—the diploid termites, relatives of cockroaches, who at least include both males and females in their worker castes.)

This explanation of an old mystery has so intrigued biologists that a subtle reversal of causality has crept into some accounts. The very existence of haplodiploidy is linked with force and elegance to the evolution of sociality, and we are almost led to believe that this mode of sex determination arose "for," or at least in the context of, the marvelous social organization of ants and bees. Yet a moment's explicit reflection assures us that this cannot be so, for two reasons.

First, *all* hymenopterans are haplodiploid, but only a few lineages within the group have developed complex social systems (most hymenopterans are asocial or minimally social wasps). The common ancestor of living hymenopterans must have been haplodiploid, but it was certainly not fully social, since the complex society of highly derived bees and ants has evolved as a phyletic afterthought in several independent lineages. Causality must run in the other direction. Haplodiploidy does not exist "for" sociality unless the future can control the past. Rather, haplodiploidy arose for other reasons and then permitted, by good and unplanned fortune, the later evolution of this wonderfully complex and successful mode of sociality. But what other reasons?—which brings me, finally, to the point of this essay, to the main reason for my fascination with *Histiostoma murchiei,* and, more immediately, to the second item.

Second, when we consider the usual ecological context of haplodiploidy in a broad range of animals that may have evolved it directly (and not merely co-opted it for another use), an interesting pattern emerges. *Histiostoma murchiei* shares a mode of life with the mites that die before birth, and with many other haplodiploid animals in distantly related groups: all are "colonizers," species that survive by seeking rare but rich resources and then reproducing as fast as they can when uncommon fortune rewards their search (the vast majority of *Histiostoma*'s hypopi die before finding a fresh earthworm cocoon). Haplodiploidy provides several advantages in this chancy approach to survival. Successful colonization does not require two separate migrations of a male and a female, or even that a single migrating female be fertilized before her search for a new resource begins. Any unmated female, even a juvenile, becomes a potential source of new colonies, since she can make a generation of males all by herself and then mate with them to begin a generation of females—the strategy evolved by *Histiostoma*.

When colonizers find a rich but ephemeral resource, haplodiploidy may enhance the speed of raising new generations by permitting fertilized females to control the sex ratio of their offspring. As I argued in my essay on "death before birth" (see *The Panda's Thumb*), when brothers mate with sisters, more offspring will populate the next generation if mothers can put most of their limited reproductive energy into making females and produce only a minimal number of males (one will often do). One male may fertilize many females, and the available number of eggs, not sperm, limits the reproductive rate of a population—so why make vast numbers of superfluous males. The principle is fine in theory, but most animals cannot easily control the sex ratio of their offspring. Despite prayers and entreaties for boys in many sexist human societies, girls continue to assert their birthright (and birth rate) of nearly 50 percent.

But many haplodiploids can control the sex ratio of their offspring. If females store sperm within their bodies after mating, any eggs that bypass the storage area become males, while those that contact it become females. Haplodiploid mites with highly unequal sex ratios often produce a brood of female eggs and then shut off the sperm supply to add a male or two right at the end.

This complex of associated features—a colonizing life style, rare and ephemeral resources, rapid reproduction, and ease of rearing new generations in strange places—seems to define the original context of advantage for haplodiploidy. If we assume, as a hypothesis only, that haplodiploidy usually arises as an adaptation for life in this uncertain world, then it must be interpreted as a lucky accident with respect to its later utility in the evolution of sociality in ants and bees.

Now what could be more different, in our usual biological thinking, than the chancy life of a solitary female colonizer (whose offspring can hardly become social on a resource that doesn't last more than a generation or two), and the complexity, stability, and organization of ant and bee societies. Is it not peculiar in the extreme that haplodiploidy, a virtual prerequisite for the evolution of hymenopteran societies, probably first evolved as an adaptation for a life style almost diametrically opposed (at least in its metaphorical implications)? If I can convince you that it is not peculiar at all, but an example of a basic principle that distinguishes evolutionary biology from a common stereotype about science in general, then this essay has succeeded.

It is a clear, though lamentably common, error to assume that the current utility of a feature permits an inference about the reasons for its evolutionary origin. Current utility and historical origin are different subjects. Any feature, regardless of how or why it first evolved, becomes available for co-optation to other roles, often strikingly different. Complex features are bursting with potentialities; their conceivable use is not confined to their original function (I confess that I have used a credit card to force a door). And these evolutionary shifts in function can be as quirky and unpredictable as the potentials of complexity are vast. It happens all the time; it virtually defines the wondrous indefiniteness of evolution.

The balancing fins of fish became the propulsive limbs of terrestrial vertebrates, while the propulsive tail became an organ that often aids in balance. The bone that suspended an ancestral fish's upper jaw to its cranium became the bone that transmits sound to the ears of reptiles. Two bones that articulated the jaws of that reptile then became the other two sound-transmitting bones of the mammalian middle ear. When we see how beautifully our hammer, anvil, and stirrup function in hearing, who would imagine that one bone once suspended jaw to cranium, while two others articulated the jaws. (By the way, before jaws even evolved, all these bones supported the gill arches of an ancestral jawless fish.) And a mode of sex determination that may first have aided a lonely female colonizer apparently became the basis of social systems rivaled only by our own in complexity.

As we probe deeper and further back, the unpredictabilities mount. I discussed the quirkiness of a functional shift toward support of sociality by a sexual system which probably evolved as an aid to colonization. But what about the larger reason for our imperfect and unpredictable world: structural limits imposed by features evolved for other reasons? Social systems, like those of ants and bees, might be of enormous advantage to hosts of other creatures. But they do not evolve largely because it is so difficult to get them started in diploid organisms (only termites have succeeded), while haplodiploid hymenopterans develop them again and again. And going one step further back (I promise to stop here), what about constraints on the evolution of haplodiploidy itself. Haplodiploidy might be a wonderful adaptation to a host of ecologies, but it cannot always be easily evolved.

Assuming that haplodiploids generally arise from diploids, what does it take to turn a haploid creature into a male? Under some systems of diploid sex determination, male haploids cannot easily evolve. A haploid human would not be male, for a single X chromosome induces the development of a sterile female. But other diploids have a so-called XX-XO system of sex determination, where females have two X chromosomes and males have a single X with no accompanying Y (but all other chromosomes in pairs). In such systems, a haploid organism might develop easily and directly into a male. (The XX-XO system is not a prerequisite for haplodiploidy, since more complex modifications can produce male haploids from other modes of diploid sex determination.)

In short, modes of sex determination limit haplodiploidy, haplodiploidy limits sociality, and sociality requires a quirky shift in the adaptive significance of haplodiploidy. What order can we find in evolution amidst such a crazy-quilt of limits to a sensibly perfect and predictable world?

Some might be tempted to read an almost mystical message into this theme—that evolution imposes an ineffable unknowability upon nature. I would strongly reject such an implication: knowledge and prediction are different phenomena. Others might try to read a sad or pessimistic message—that evolution isn't a very advanced science, or isn't even a science at all, if it can't predict the course of an imperfect world. Again, I would reject any such reading of my words about constraint and quirky functional shift.

The problem lies with our simplistic and stereotyped view of science as a monolithic phenomenon based on regularity, repetition, and ability to predict the future. Sciences that deal with objects less complex and less historically bound than life may follow this formula. Hydrogen and oxygen, mixed in a certain way, make water today, made water billions of years ago, and presumably will make water for a long time to come. Same water, same chemical composition. No indication of time, no constraints imposed by a history of previous change.

Organisms, on the other hand, are directed and limited by their past. They must remain imperfect in their form and function, and to that extent unpredictable since they are not optimal machines. We cannot know their future with certainty, if only because a myriad of quirky functional shifts lie within the capacity of any feature, however well adapted to a present role.

The science of complex historical objects is a different, not a lesser, enterprise. It seeks to explain the past, not predict the future. It searches for principles and regularities underlying the uniqueness of each species and interaction, while treasuring that irreducible uniqueness and describing all its glory. Notions of science must bend (and expand) to accommodate life. The art of the soluble, Peter Medawar's definition of science, must not become shortsighted, for life is long.

2 | Personalities

5 The Titular Bishop of Titiopolis

MODERN GEOLOGY BEGAN, or so the usual story goes, with the publication of a book so oddly named that it almost surpasses the peculiarity of the title later assumed by its author, Nicolaus Steno, a Dane by birth and a Catholic convert who became Titular Bishop of Titiopolis *(in partibus infidelium)* in 1677. (Titular bishops "preside" over areas in pagan hands and therefore unavailable for actual residence—in the realm of infidels, as the Latin subtitle proclaims. The old bishopric of Titiopolis is now part of Turkey.) As his real job, dangerous enough in Protestant lands, Steno ministered to the scattered Catholic remnants of northern Germany, Norway, and Denmark.

The book, published in 1669, bears a title considered "almost unintelligible" by its chief translator from the original Latin. It is called *De solido intra solidum naturaliter contento dissertationis prodromus,* or *Prodromus to a dissertation on a solid body naturally contained within a solid.* A prodromus is an introductory discourse, but Steno never wrote the promised dissertation because his religious interests, following his conversion in 1667 and his ordination in 1675, led him to abandon his distinguished career as a medical anatomist and, by fortuitous introduction at the very end of his scientific work, a geologist.

Why a solid within a solid? And what can such a cryptic phrase have to do with the origin of modern geology? Posing a problem in a startling and novel way is the virtual prerequisite of great science. Steno's genius lay in recognizing that a solution to the general problem of how solid bodies get inside other solids might provide a criterion for unraveling the earth's structure and history. But Steno did not formulate his problem by rational deduction from his armchair. As so often happens in a human world, he drifted toward it after an accidental beginning.

Like many anatomists, Steno became interested in the resemblances of humans with other animals. He decided to dissect sharks and made some important discoveries. He demonstrated, for example, that the tight coils of the spiral intestine yielded the same total length (within a more confined space) as the meandering intestine of mammals. In October 1666, during Newton's great year, or *annus mirabilis,* and a month after London burned, Steno received for study the head of a giant shark caught at the city whose English name, Leghorn, is as peculiar as Steno's two titles. (The name refers neither to limbs nor musical instruments, but represents a poor English rendering of the old spelling, Ligorno, for the town now called Livorno in Italian.) Steno, like so many intellectuals, was working at the nearby city of Florence under the patronage of Ferdinand II, the Medici grand duke. In examining the teeth of his quarry, Steno recognized that he had accidentally bought into one of the major scientific debates of his age, the origin of *glossipetrae,* or tongue stones.

These fossil sharks' teeth could be collected by the barrel, especially in Malta. In twentieth-century terms, their origin cannot be doubted. They are identical with the teeth of modern sharks in outward form, detailed structure, and chemical composition—therefore they cannot be anything but sharks' teeth.

Yet the identity in form that makes us so certain led to another potential interpretation in Steno's time—for God, the author of all things, often created with striking similarity in different realms to display the order of his thoughts and the glorious harmony of his world. If he had made a world with seven planets (sun, moon, and the five visible planets of an older cosmology) and seven notes in a musical scale,

A mid-eighteenth century illustration of why *glossipetrae* (A, B, and C) must come from the mouths of sharks. FROM DE CORPORIBUS MARINIS LAPIDESCENTIBUS (ON PETRIFIED MARINE BODIES) BY THE SICILIAN ARTIST-SCIENTIST AUGUSTINO SCILLA.

why not imbue rocks with the plastic power to form objects precisely like the parts of animals? After all, the *glossipetrae* came from rocks and rocks were created as we find them. If the tongue stones are sharks' teeth, how did they get *inside* rocks? Moreover, the earth is only a few thousand years old, and tongue stones inundated European collections. How many sharks could have infested Mediterranean waters in so short a time?

Steno observed that his shark had hundreds of teeth and that new ones formed continually as old teeth wore down and fell out. The numbers of *glossipetrae* from Malta no longer foreclosed an origin in sharks' mouths, even under the Mosaic chronology (which Steno did not question). According to the common legend that great scientists are unprejudiced observers who can shuck constraints of culture and see nature directly, Steno came to his correct conclusion—that *glossipetrae* are fossil sharks' teeth—because he made better observations. Steno was a fine observer, but he was also an adherent to the new mechanical philosophy that insisted on physical causes for phenomena and viewed detailed internal similarity as a sure sign of common manufacture in the mechanical sense. Steno did not see better; rather, he possessed the conceptual tools to interpret his excellent observations in a necessary way that we continue to regard as true.

But Steno then abstracted the problem of *glossipetrae* in a remarkably original manner—and achieved with this great insight his role as the founder of modern geology. The tongue stones found within rocks, Steno reasoned, were problematic because they were solids enclosed within a solid body. How did they get in there? Steno then recognized that all the troubling objects of geology were solids within solids—fossils in strata, crystals in rocks, even strata themselves in basins of deposition. A general theory for the origin of solids within solids could provide a guide for understanding the earth's history.

Taxonomy is often regarded as the dullest of subjects, fit only for mindless ordering and sometimes denigrated within science as mere "stamp collecting" (a designation that this former philatelist deeply resents). If systems of classification were neutral hat racks for hanging the facts of the world, this disdain might be justified. But classifications both reflect and direct our thinking. The way we order represents the way we think. Historical changes in classification are the fossilized indicators of conceptual revolutions.

The French scholar Michel Foucault uses this principle as his key for understanding the history of thought. In *Madness and Civilization,* for example, he notes that a new method of dealing with the insane arose in the mid-seventeenth century and spread rapidly throughout Europe. Previously, madmen had been exiled or tolerated and allowed to wander about. In the mid-seventeenth century, they were confined in institutions along with the indigent and unemployed, a motley assemblage by modern standards. We might regard this classification as senseless or cruel, but as

Foucault argues, such a judgment will not help us to understand the seventeenth century.

Why classify together the poor, the unemployed, and the insane; what common theme could inspire such an ordering? Foucault argues that the birth of modern commercial society led to a new designation of the cardinal sin, the one that had to be made invisible by confining all those who, for whatever reason, wallowed in it. That sin was idleness, and Foucault shows that sloth replaced the old medieval curse of pride as the most fundamental of the seven deadly sins in seventeenth-century texts. It mattered little that the insane did not work for biological or psychological reasons, and the unemployed for want of opportunity.

Steno also reordered the world in a way that must have seemed as curious to his contemporaries as the amalgamation of madness and poverty seems to us. As his contemporaries gathered the idle, Steno identified solids within solids as a fundamental class of objects, divided them from everything else, and developed a set of criteria to sort his solids into subdivisions representing the different causes that fashioned them. The great *Prodromus* is, fundamentally, a treatise on a new system of classification for solids within solids—a classification by common genesis, rather than superficial similarity of outward appearance. Steno's revolution in thought arises from his altered classification—and his curious title, so understood, could not be more devastatingly appropriate. I have read the *Prodromus* many times, but when I finally understood its message, just last month, that bizarre title sent a shiver up my spine.

The *Prodromus* has usually been misinterpreted by geologists who attribute Steno's success to his use of modern observational methods. (In fact, although the *Prodromus* is sprinkled with astute observations, its longest section is a speculative discussion on the origin of solid bodies, based on the incorrect premise that all solids must be generated from liquids, and that the form of a solid indicates the motions of the liquids that produced it.) His translator writes, for example: "At a time when fantastic metaphysics were rife, Steno trusted only to induction based upon experiment and observation." But the *Prodromus* reports no real experiments and only a modest number of observations. It succeeded primarily because Steno followed a metaphysic congenial with our own, but relatively new in his time.

Geologists have also judged Steno inappropriately by searching the text for gems of "modern" insight, rather than by understanding its argument as a totality. Thus, the commonest statement about the *Prodromus,* often the only statement made by geologists, holds that Steno presented the crystallographic law of the constancy of interfacial angles—that however much the size and shape of crystal faces vary, the angles between them remain the same. Well, perhaps he did, but the "law" appears as two throwaway lines in a figure caption, and has little relation to Steno's major theme or argument. (It arises simply as a corollary to his speculations about inferring the motion of fluids from the form of solids precipitated from them.)

No, the *Prodromus* is, as its title states, about solids in solids and their proper classification by mode of origin. It is founded upon two great taxonomic insights: first, the basic recognition of solids within solids as a coherent category for study and, second, the establishment of subdivisions to arrange solids within solids according to the causes that fashioned them.

Steno uses two criteria for his subdivisions. (They are blessedly obvious once you state the problem, but Steno's revolution is the statement itself.) First, in what might be called the principle of molding, Steno argues that when one solid lies within another, we can tell which hardened first by noting the impress of one object upon the other. Thus, fossil shells were solid before the strata that entomb them because shells press their form into surrounding sediments just as we make footprints in wet sand. But surrounding rocks were solid before the calcite veins that run through them because the calcite fills preexisting channelways just as Jello matches the flutes of a mold. The principle of molding allows us to establish the temporal order of formation for two objects in contact. In a world still regarded by many of Steno's contemporaries as formed all at once by divine fiat, this criterion of history struck a jarring chord and eventually forced a transposition in thought.

Early in the *Prodromus,* Steno stated the problem that he wished to solve with his second criterion: "Given a substance possessed of a certain figure, and produced according to the laws of nature, to find in the substance itself evidences disclosing the place and manner of its production." His solution, the basic principle for any historical reconstruction, holds:

> If a solid substance is in every way like another solid substance, not only as regards the conditions of its surface, but also as regards the inner arrangement of parts and particles, it will also be like it as regards the manner and place of production.

Past processes cannot be observed in principle; only their results remain. If we wish to infer the processes that formed any geological object, we must find clues in the object itself. The surest clue is detailed similarity—part by internal part—with modern objects formed by processes we can observe directly. Similarity can be misleading—and great mistakes have been made in applying Steno's principle—but our confidence in common origin mounts as we catalog more and more detailed similarities involving internal structure and chemical composition as well as external form.

Thus, Steno argues, sedimentary rocks must be the deposits of rivers, lakes, and oceans because they "agree with those strata which turbid water deposits." Fossil shells once belonged to animals, and crystals precipitated from fluids as we make salt or rock candy today.

With these two principles—molding and sufficient similarity—Steno established both prerequisites for geological, or any historical, reconstruction: he could determine how and where objects formed, and he could order events in time. Steno's genius, to say it one more time, lay in establishing this new conceptual framework for observation, not in the acuity of the subsequent observations themselves. Steno's break with older traditions stands out most clearly in his complete failure, save in one sheepish passage, to consider the primary subject that obsessed his colleagues: the identification of goals and purposes for all things, including what we now regard as purely physical processes of uplift, erosion, and crystallization. In one fleeting passage, Steno cites dissimilarity of function as a reason subservient to his usual argument about internal resemblance for stating that rocks and bones form differently. But he quickly adds the disclaimer, "if one may be permitted to affirm aught about a subject otherwise so little known as are the functions of things." As Foucault also argues, the subjects you leave out of your taxonomies are as significant as the ones you put in.

The four-part organization of the *Prodromus* has generally been viewed as disjointed or even incoherent—thrown together by a man itching to leave Florence but forced to justify the grand duke's patronage. I view it instead as a comprehensive and tightly reasoned brief for a science of geology based upon the two principles of molding and sufficient similarity.

Part one is a teaser, a specific example to demonstrate the power of the general method. The *glossipetrae,* Steno argues, must be sharks' teeth because they are identical in form and internal arrangement with the objects he had plucked from the mouth of his quarry from Leghorn. They solidified before the rocks that enclose them because they impress their form upon the surrounding sediment. Therefore—and now the argument begins to move toward revolutionary generality—sedimentary rocks were not created with the earth, but have formed as the deposits of turbid waters in rivers, lakes, or oceans. Moreover, similar marine fossils are often found high in mountains and far from the sea; these fossils also solidified before the strata enclosing them. Thus, the earth has an extensive history: seas and lands have changed places, and mountains have emerged from the waters.

In the second part, Steno argues that *glossipetrae* are but one example of the general problem of solids within solids, and that the principles of molding and sufficient similarity can establish proper taxonomic subdivisions based on common modes of origin. The third part treats the major classes of solids within solids and establishes two basic categories for objects within rocks: fossils that harden *before* the enclosing strata, and crystals and veins that form *within* solid rocks.

The fourth part, a reconstruction of the geological history of Tuscany, has been problematic or even embarrassing to geologists who wish to view Steno as their founding saint. (The Catholic Church, by the way, is also considering Steno for sainthood, and he may eventually attain an unprecedented double distinction.) Steno constructs his history to match biblical chronology, with two cycles of deposition—from the original void and from Noah's universal ocean. The essence of this part, however, is not his continued loyalty to Moses—Steno was, after all, not a man of our century—but rather his demonstration that the principles of molding and sufficient similarity can be used not only to classify objects (part three), but also to reconstruct the history of the earth from these objects (part four). The last part of the *Prodromus* demonstrates by specific example, drawn from the local terrain, that the proper classification of solids within solids can establish a science of geology.

In 1678, Athanasius Kircher published a figure showing all letters of the alphabet, including the contraction Æ etched in veins of calcite. Today, we chuckle and dismiss the well-formed letters as accidents. But to Kircher they were no less significant than the shells of clams also found in rocks. One might argue that clams are more complex than letters, but a Venetian work of 1708 depicted an agate that seemed to show, in its bands of color, Christ on the cross with all proper accouterments, including a sun on the favored right side and a moon on the despised left. The caption proclaimed in German doggerel: *Solche wunderbarliche Gestalt, hat die Natur in ein Agat gemahlt*—"Nature herself has painted this wonderful figure in an agate." Why was a clam in a rock different from a letter or a crucifixion? Since alphabets and religious scenes cannot be preexisting objects buried in strata, they must be made by a plastic power in the rocks themselves. As long as "odd things in rocks" formed a single category, clams and sharks' teeth would also be manifestations of the plastic force and no science of paleontology or of historical geology would be possible. But Steno's classification recognized the basic distinction between fossils that hardened before the rocks that enclosed them and intruding veins that might by accident resemble some abstract form or design.

Steno changed the world in the simplest and yet most profound way. He classified its objects differently.

6 | Hutton's Purpose

IN HIS TRIBUTE to Lucretius, Virgil wrote: "Happy is he who could learn the causes of things" *(Felix qui potuit rerum cognoscere causas).* A noble and uncomplicated sentiment to be sure, but an even more illustrious predecessor had shown that causality is no simple matter. Aristotle, in the *Posterior Analytics* of the *Organon,* stated: "We only think that we have knowledge of a thing when we know its cause." He then proceeded to give a complex analysis of the concept of causality itself.

Each event, Aristotle argued, has four distinct kinds of causes. Consider the so-called parable of the house, the standard example, probably in continual use for more than two thousand years, for illustrating Aristotle's schema. What is the cause of my house? What are the *sine quibus non,* the various factors whose absence would lead to no house at all or to a house of markedly different design?

First, Aristotle argues, we must have the straw, sticks, or bricks—the *material* cause. It obviously matters, as the three little pigs discovered, what material you choose. Next, someone must do the actual work, thatch the roof or lay the bricks—the effector, or *efficient* cause. The blueprint that the mason follows doesn't do anything actively, and it is not building material. But it is a cause of sorts, since different blueprints yield different houses and no plan at all leaves you with a pile of bricks. These preconceived marching orders are *formal* causes in Aristotle's lexicon. Finally, if the house served no purpose as an abode for its inhabitants, no one would bother to build it. Purposes are *final* causes.

We do not follow Aristotle's analysis in our linguistic habits today; our entire notion of "cause" is now pretty much restricted to Aristotle's efficient causes. We do not deny the material and formal aspects, but we no longer call them causes. When I identify the motion of my pool cue as *the* cause of a ball's errant trajectory (though only an efficient cause to Aristotle), I do not regard the composition of the ball or the blueprint of the table as irrelevant, but I no longer call them causes.

The elimination of purpose, or final cause, tells a more important story and represents a major change in style for Western science. Aristotle saw nothing absurd in granting each event both an efficient cause (a mechanism, in our terminology) and a final cause (a purpose). He writes, for example,

> Light shines through a lantern. Being composed of particles smaller than the pores of the lantern, it cannot help passing through them (assuming that this is how the light is propagated); but it also shines for a purpose, so that we may not stumble [*Posterior Analytics,* 94b, 1. 28].

We can follow Aristotle for devices constructed by humans for definite purposes. We did put holes in lanterns to let the light through. Final cause also remains a legitimate concept for the adaptations of organisms, even though these features arise by natural processes and not by any conscious activity of the animals involved. It remains good vernacular English to say that bats and birds have wings "for" flight, and the wolf legitimately invoked final cause in replying to Red Riding Hood's inquiry about the sharpness of his teeth, "All the better to eat you with, my dear."

But we balk at ascribing final causes to the physical workings of inanimate objects, although Aristotle did not. Aristotle was comfortable with the idea that "it thunders both because there must be a hissing and roaring as the fire is extinguished, and also (as the Pythagoreans hold) to threaten the souls in Tartarus and make them fear" [*Ibid.*, 94b, l. 34]. We chuckle at Aristotle here, and that chuckling represents perhaps the greatest change that science has undergone in modern times. We no longer view the universe as explicitly designed in all its minor and multifarious parts to serve some human purpose. We have replaced this cosmic hubris with a more mechanical view of nature. God may have wound the clock and established the laws of ticking at the outset, but he surely does not spend his precious time fashioning each blade of grass and grain of sand to provide explicit instruction or sustenance for his favored species on earth. The mechanical view, based on the primacy of efficient causation, has properly banished final cause from the domain of natural, physical objects.

So absolute is this proscription of final cause, and so essential to a modern definition of science, that old passages about the final causes of physical objects are unsurpassed as targets of ridicule when, in our arrogant approach to history, we choose to flay the past, all the better to bask in our current wisdom (a legitimate final cause in human psychology). It would, indeed, be hard to deny that many of these passages are, well, simply funny.

Louis Agassiz, for example, seriously argued in the 1860s that ice ages could be understood both by the physics of glacial motion (efficient cause) and as a dispensation of divine benevolence designed to churn and enrich the soil:

> One naturally asks, What was the use of this great engine set at work ages ago to grind, furrow, and knead over, as it were, the surface of the earth? We have our answer in the fertile soil which spreads over the temperate regions of the globe. The glacier was God's great plough.

In 1836, William Buckland, Oxford's first academic geologist, claimed that abundant coal, the fuel of England's glory, was so cleverly distributed in the bowels of the earth that God himself must have placed it there, many million years ago, in loving preparation for its future use. We might rightly suspect that as old rock, coal should now be buried under so many miles of younger strata that it would lie beyond (or, rather, beneath) our reach. But God saw fit to ordain its deposition not in vast horizontal sheets but in discontinuous bowl-shaped basins whose edges still intersect or lie just a bit below the earth's surface. Moreover, the strata that were buried at inaccessible depths have often been extensively faulted and uplifted to the surface. These faults are a further boon to miners because the streams that often run along their fractured boundaries can guide us to the precious substance underneath, and because destructive fires can no longer ravage an entire field when faults break an extensive stratum into several discontinuous segments separated by rock that will not burn. Buckland wrote:

> However remote may have been the periods, at which these materials of future beneficial dispensations were laid up in store, we may fairly assume that . . . an ulterior prospective view of the future uses of Man formed part of the design, with which they were, ages ago, disposed in a manner so admirably adapted to the benefit of the Human Race.

Alexander Winchell, prominent American geologist and first chancellor of Syracuse University, could scarcely contain himself in paying lyrical tribute (in *Sketches of Creation,* 1870) to the faults that bring coal within our orbit:

> Buried ten thousand feet from view, man would never have learned of its existence, much less would he have known how to raise it to the surface. See the provision of Nature in breaking up the coal-bearing strata and tilting them on edge, as much as to say, "Lo! here is your desire; search not in vain; dig, and be satisfied with warmth; drive forth the hidden energy . . . and bid

the servants furnished to your hands execute all the behests of your convenience."

I have, needless to say, no desire to resurrect this tradition of argument about final causes. But I would condemn, for two reasons, any attempt to parade old passages about final cause as a source for ready laughter and self-congratulation in transcending past ineptitude. First, it subverts any effort to understand the past and use it as a guide for interpreting the present. When such fine intellects as Agassiz and Buckland (not to mention Aristotle) seriously advance these arguments, we must view them as markers of a fundamentally different conception of the world, not as signs of personal stupidity or general naïveté. Second, even failed views of the world, when they have both grandeur and depth, can serve as wonderfully fruitful sources of insight. To recycle a favorite quote used in my previous volume, *The Panda's Thumb,* "Give me a fruitful error any time, full of seeds, bursting with its own corrections. You can keep your sterile truth for yourself" (Pareto's comment on Kepler).

Final cause served as such a fruitful error at a crucial moment in my own profession. The greatest reconstruction of geology—James Hutton's theory of the earth—rested squarely on an argument about final causes. And few geologists have the slightest inkling of this "antiquated" wellspring of insight because it has been subverted by a comfortable (and comforting) myth that locates the source of Hutton's success in his pursuit of modern tactics—fieldwork and a mechanical concept of physical causality (see G.L. Davies, *The Earth in Decay* [American Elsevier, 1969], for a fine account of this myth and its correction).

Born in Edinburgh in 1726, James Hutton hobnobbed with the likes of Adam Smith and James Watt in the great Scottish intellectual circle that so influenced the life of eighteenth-century Europe. After abandoning an apprenticeship in law, he studied medicine. As a man of means, he felt no need to practice and opted instead for farming (on land inherited from his father and after bolstering his economic security by inventing and marketing a process for producing sal ammoniac from chimney soot). No rustic he, but no gentleman farmer either, he studied the latest methods in husbandry and ran a profitable, modern, model farm. In his early forties, Hutton gave up country life, returned to Edinburgh and spent the remaining three decades of his life as a full-time unemployed, intellectual gentleman.

In 1788, Hutton published his reconstruction of geology in the first volume of the *Transactions of the Royal Society of Edinburgh* (expanded in 1795 to a multivolumed work entitled *Theory of the Earth,* following a blistering attack for supposed atheism and other improprieties by the Irish chemist and mineralogist Richard Kirwan). Hutton, although usually cast as a modern empiricist, really belongs to the great tradition of comprehensive (and at least partly speculative) system building that dominated most of eighteenth-century "geology" (the term had not yet been invented, and no profession, with formally recognized procedures, then existed).

Hutton's system, his "world machine," embodied a cyclical notion of history—dynamic and endlessly recurring, but moving nowhere, as the Preacher of Ecclesiastes proclaimed:

> All the rivers run into the sea; yet the sea is not full. Unto the place from whence the rivers come, thither they return again [1:7]. The thing that hath been, it is that which shall be; and that which is done, is that which shall be done; and there is no new thing under the sun [1:9].

This view of history contrasted sharply with the more familiar Christian concept of a linear and directional sequence moving ever onward from creation to resurrection. (Hutton, who was decidedly not an atheist, did not deny that God had ordained a beginning and would decree an end. But these miraculous events lie outside the purview of science. In between these singularities, God rested and permitted the world to run by the natural laws that he had established. Only this period could be studied by science, and here Hutton discerned no direction, but only endless cycling.)

Hutton's theory rests, in part, on his choice of metaphors for the earth. With friendship for James Watt and with reverence for Isaac Newton, Hutton chose to see the world as a perfect machine that, once wound, would run forever (or until God changed the rules) without wearing out or breaking down. "This world," Hutton proclaimed,

> is an active scene or a material machine moving in all its parts. We must see how this machine is so contrived, as either to have those parts to move without wearing or decay, or to have those parts, which are wasting and decaying, again repaired.

Hutton therefore contrived a four-stage, cyclical theory of earth history. In the first stage, the only one we can observe directly, the land is worn away by erosion and eventually (stage two) deposited as strata in the depths of the ocean. There (stage three), the strata are compacted and consolidated by heat (both from the interior fires of the earth and from the weight of overlying sediments), and then (stage four), as a result of the same internal heat, fractured and uplifted to form new continents. Land and sea have changed places and the cycle starts again: erosion, deposition, consolidation, and uplift—forever and ever. Thus, in the most famous words ever written by a geologist, Hutton ends his 1788 treatise by explicitly comparing his world machine with the endless cycling of planets about the sun:

> Having, in the natural history of this earth, seen a succession of worlds, we may from this conclude that there is a system in nature; in like manner as, from seeing revolutions of the planets, it is concluded, that there is a system by which they are intended to continue those revolutions. . . . The result, therefore, of our present enquiry is, that we find no vestige of a beginning,—no prospect of an end.

Although Hutton had predecessors for each individual claim, the revolutionary content of his comprehensive system has two primary sources. First, he burst the boundaries of time, thereby establishing geology's most distinctive and transforming contribution to human thought—"deep time," as John McPhee *(Basin and Range*, 1981*)* puts it. In his next most famous statement, Hutton wrote: "Time, which measures everything in our idea, and is often deficient to our schemes, is to nature endless and as nothing."

One of the major barriers to an acceptance of deep time had been the absence of any recognized restoring force in the operations of nature. Geologists before Hutton generally lacked a "concept of repair." They knew that erosion constantly wore down the land, and cultural traditions supported the idea of history as a continual decline from the original perfection of Eden. They did not recognize that the earth's internal heat could fracture and raise vast areas of the continents into mountains and high plains; (they regarded mountain ranges as part of the earth's original structure and volcanoes as mere pimples on the globe's degrading surface). Without such a concept of repair, the earth must be very young. After all, it would not take long to erode all continents beneath the sea, and mountains still tower above us. As his second great contribution, Hutton demonstrated that igneous forces within the earth supplied a restorative power that uplifted continents, prevented the land's destruction, permitted a theory of endless cycling, and established the possibility of deep time.

For each of these contributions—deep time and a concept of repair—Hutton supplied a key empirical observation. For time, Hutton recognized the significance of what geologists call an angular unconformity. Old sedimentary rocks, originally deposited in horizontal sheets, are often uplifted and tilted during the operation of Hutton's restorative forces. They may then be eroded down, covered again by water, and overlain by a new sequence of horizontal sediments. The contact between these two packages of sedimentary

Hutton's original figure of an angular unconformity. Note vertical strata below and horizontal above. DRAWING BY JOHN CLARK APPEARED IN HUTTON'S 1795 TREATISE.

rocks is called an angular unconformity because the tilted older strata meet the horizontal younger strata at an angle. Hutton rejoiced in these unconformities because they yielded direct evidence for his theory of cycles. Each angular unconformity recorded *two* Huttonian worlds placed in sequence one atop the other—an older world in the first package, made in the depths of the sea, uplifted, and eroded down again, and a younger world in the second package, made in a later ocean and now uplifted to our view. John Playfair, Hutton's greatest interpreter and the most literate man who ever wrote about geology, recorded his awe upon viewing an angular unconformity on a field trip with Hutton:

> What clearer evidence could we have had of the different formation of these rocks, and of the long interval which separated their formation, had we actually seen them emerging from the bosom of the deep? . . . The

> mind seemed to grow giddy by looking so far into the abyss of time.

For a concept of repair, Hutton recognized the igneous nature of two common rocks, basalt and granite. Many geologists at the time argued that basalt and granite were sedimentary rocks, deposited from water; Hutton held (correctly) that they had risen as magma from the depths of the earth and cooled to their present state. Thus, they represented the products of Hutton's restorative force. This issue became the focus of a great struggle in science, the debate between Neptunists, who advocated water, and Plutonists (like Hutton), who opted for internal fires as the source of granite and basalt. The debate received a good popular press and even spilled onto the pages of *Faust* (Goethe being, among other things, a brilliant geologist) where, from the error of its author, Faust argues for water and Mephistopheles (only appropriately) for fires within the earth. Arcane scientific debates don't receive this much notice unless the stakes are high—and indeed they were. Basalt and granite occupy vast areas of the earth's surface. If they, like all other common rocks then recognized, are sedimentary, then all rocks may be products of an original ocean, and the entire history of our earth may be short and directional—a few thousand years of deposition and drying out. But if granite and basalt are igneous, then they record a restorative force of sufficient power to cover much of the earth with its products. History may be cyclical and long. Hutton relied primarily upon field evidence for his Plutonian conclusions. He noted, in particular, that granite and basalt often occur as vertical dikes cutting through horizontal sediments and marking the passageway of magmas from the earth's interior.

Did Hutton base his general theory upon these observations? Did he triumph, as the usual story goes, because he was an objective modernist who combated ancient traditions of prejudiced speculation by using the "real" scientist's tool of pure and unfettered observation, and by holding a modern concept of mechanical causality? Hutton's countryman, the great Scottish geologist Sir Andrew Geikie, gave this common myth its strongest support in his 1905 volume, *The Founders of Geology.* Geikie wrote: "In the whole of Hutton's doctrine he rigorously guarded himself against the admission of any principle which could not be founded on observation. He made no assumptions. Every step in his deductions was based upon actual fact." Geikie's heroic Hutton gathered his facts by the method that provides both the strength and the mystique of geology—fieldwork:

> He went far afield in search of facts, and to test his interpretation of them. He made journeys into different parts of Scotland. . . . He extended his excursions likewise into England and Wales. For about thirty years, he had never ceased to study the natural history of the globe, constantly seeking to recognize the proofs of ancient terrestrial revolutions, and to learn by what causes they had been produced.

This Hutton matches the idealized image of geology presented to generations of students, but it bears little relation to the original. To be sure, Hutton did not remain perennially in his armchair. He made many excursions and saw many things. His observations no doubt inspired and instructed him; but we can show, also without doubt, that fieldwork was not the source of his theory. For his two key observations, the chronology of the official myth is backward. Hutton saw his first angular unconformity after he had presented his full-blown theory in public. Moreover, by his own admission, he had observed granite in only one uninformative place before publishing his theory. Fieldwork, at best, provided confirmation for a theory developed elsewhere.

When we consult Hutton's written record, we find—if we may take his own presentation at face value—that he developed his general theory by the accepted route of eighteenth-century system builders: he reasoned from his own version of first principles and then gathered arguments for what he regarded as necessary conclusions. And when we examine Hutton's concept of first principles, we find that he was not a mechanist committed to empirical test, but a follower of Aristotle's notion of causality.

Hutton did have a mechanical concept of causality; his earth is a perfect machine, working with no hint of senescence until God chooses to ordain an end. But Hutton followed Aristotle in arguing that events have *both* a mechanical (or efficient) cause and a purpose, or final cause. Of the two, Hutton clearly regarded final causes as more important and more fundamental to his system. When Geikie and others chose to ignore Hutton's own writing, and to use him as a moral homily for an idealized view of science, they did major disservice to a great, if not a modern, intellect.

The very first paragraph of Hutton's great work (the original 1788 version), in emphasizing both machines and purposes, advances the Aristotelian argument that any adequate theory of the earth must explain both how and why:

> When we trace the parts of which this terrestrial system is composed, and when we view the general connection of those several parts, the whole presents a machine of a peculiar construction by which it is adapted to a certain end. We perceive a fabric, erected in wisdom, to obtain a purpose worthy of the power that is apparent in the production of it.

In the fourth paragraph, we learn that the earth's final cause must be expressed in terms of fitness for its sentient inhabitants, namely us: "This globe of the earth is a habitable world; and on its fitness for this purpose, our sense of wisdom in its formation must depend."

Hutton then explains how he developed his general theory of the earth as a self-restoring machine with a cyclical history of erosion, deposition, consolidation, and uplift. He appeals neither to field observations nor to mechanical causes but bases his argument on a puzzle arising from his own experience as a farmer and centered squarely on the idea of final cause. We may refer to this puzzle as the "paradox of the soil."

Without soil for agriculture, we could not support ourselves on this planet. Soil is a product of erosion, the destructive phase of the Huttonian cycle:

> A solid body of land could not have answered the purpose of a habitable world; for a soil is necessary to the growth of plants; and a soil is nothing but the materials collected from the destruction of the solid land. . . . The heights of our land are thus leveled with the shores; our fertile plains are formed from the ruins of mountains.

Now, the paradox. To form the soil so necessary for our lives and, therefore, so essential to the earth's final cause, nature uses a mechanical process that must destroy the land: "We are, therefore, to consider as inevitable the destruction of our land, so far as effected by those operations which are necessary in the purpose of the globe, considered as a habitable world." But God would not play such a joke on his favored creatures. He could not employ as a source of life-giving soil a process that must soon obliterate all humanity by washing our land into the sea. A restorative force must exist a priori, so that the earth may display wisdom in its adaptation for human life:

> If no such reproductive power, or reforming operation, after due enquiry, is to be found in the constitution of this world, we should have reason to conclude, that the system of this earth has either been intentionally made imperfect, or has not been the work of infinite power and wisdom.

Hutton did not find his restorative force unexpectedly in the field by stumbling upon an angular unconformity or pondering the nature of granite. He deduced the necessity of a restorative force from a threatening paradox in final cause, and then set out to find it. Indeed, he portrays his entire treatise as an earnest search for purpose in physical objects:

> This is the view in which we are now to examine the globe; to see if there be, in the constitution of this world, a reproductive operation, by which a ruined constitution may be again repaired, and a duration or stability thus procured to the machine, considered as a world sustaining plants and animals. . . . Here is an important question . . . a question which, perhaps, it is in the power of man's sagacity to resolve; and a question which, if satisfactorily resolved, might add some lustre to science and the human intellect.

When Hutton locates his restoring forces in the earth's internal fire, he continues the Aristotelian strategy of identifying both how they work and why, in human terms, they operate as they do:

> The end of nature in placing an internal fire or power of heat, and a force of irresistible expansion, in the body of this earth, is to consolidate the sediment collected at the bottom of the sea, and to form thereof a mass of permanent land above the level of the ocean, for the purpose of maintaining plants and animals.

Volcanoes, Hutton tells us, are "not made on purpose to frighten superstitious people into fits of piety and devotion, nor to overwhelm devoted cities with destruction." They are escape vents for internal fires, "spiracles to the subterranean furnace, in order to prevent the unnecessary elevation of land, and fatal effects of earthquakes." Some may die in their eruptions, but only so that more may live: "While it may occasionally destroy the habitations of a few, it provides for the security and quiet possession of the many."

Hutton's contemporaries certainly understood the central role of final cause in his theory, both as an original motivation and a sustaining theme. Playfair wrote of his treatise: "We see everywhere the utmost attention to discover, and the utmost disposition to admire, the instances of wise and beneficent design manifested in the structure, or economy of the world." Hutton, he continued, regarded final causes as preeminent:

> They were the parts . . . which he contemplated with greatest delight; and he would have been less flattered, by being told of the ingenuity and originality of his theory, than of the addition which it had made to our knowledge of final causes.

I am not, of course, suggesting that final cause be readmitted into science as a component for the explanation of physical events. I merely wish to point out that, although theories may be winnowed and preserved empirically, their sources are as many as people and times and traditions and cultures are varied. If we use the past only to create heroes for present purposes, we will never understand the richness of human thought or the plurality of ways of knowing.

Final cause inspired the greatest of all geological theories, but we may use it no longer for physical objects. This creative loss is part of Darwin's legacy, a welcome and fruitful retreat from the arrogant idea that some divine power made everything on earth to ease and inform our lives. The extent of this loss struck me recently when I read a passage from the work of Edward Blyth, a leading creationist of Darwin's time. He wrote of the beauty and wisdom "so well exemplified in the adaptation of the ptarmigan to the mountain top, and the mountain top to the habits of the ptarmigan." And I realized that this little line expressed the full power of what Darwin had wrought—for while we may still speak of the ptarmigan adapting to the mountain, we may no longer regard the mountain as adapted to the ptarmigan. In this loss lies all the joy and terror of our current view of life.

7 The Stinkstones of Oeningen

IN HIS MANIFESTO for a science of paleontology, Georges Cuvier compared our ignorance of geological time with our mastery of astronomical space. He wrote, in 1812, in the preliminary discourse to his great four-volume work on the bones of fossil vertebrates:

> Genius and science have burst the limits of space, and . . . have unveiled the mechanism of the universe. Would it not also be glorious for man to burst the limits of time. . . . Astronomers, no doubt, have advanced more rapidly than naturalists; and the present period, with respect to the theory of the earth, bears some resemblance to that in which some philosophers thought that the heavens were formed of polished stone, and that the moon was no larger than the Peloponnesus; but, after Anaxagoras, we have had our Copernicuses, and our Keplers, who pointed out the way to Newton; and why should not natural history also have one day its Newton? [I have followed the famous Jameson translation of 1817, which is as canonical for Cuvier's *Discours préliminaire* as its namesake King James's is for Moses—hence some pleasant archaisms throughout, although I have checked the original in all cases for accuracy.]

Cuvier, an ambitious man, may have held personal hopes, though Darwin (whose earthly remains do lie next to Newton's in Westminster Abbey) has generally commandeered the proffered title. Still, Cuvier didn't do badly. His immediate successors, at least in France, usually referred to him as the Aristotle of biology.

The centenary of Darwin's death (April 1882) has prompted a round of celebrations throughout the world. But 1982 is also the sesquicentenary of Cuvier's demise (1769–1832), and our erstwhile Aristotle has attracted scant notice. Why has Cuvier, surely the greater giant in his own day, been eclipsed (at least in the public eye) during our own? In power of intellect, and range and breadth of output, Cuvier easily matched Darwin. He virtually founded the modern sciences of paleontology and comparative anatomy and produced some of the first (and most beautiful) geological maps. Moreover, and so unlike Darwin, he was a major public and political figure, a brilliant orator, and a high official in governments ranging from revolution to restoration. Charles Lyell, the great English geologist, visited Cuvier at the height of his influence and described the order and system that yielded such a prodigious output from a single man:

> I got into Cuvier's sanctum sanctorum yesterday, and it is truly characteristic of the man. In every part it displays that extraordinary power of methodising which is the grand secret of the prodigious feats which he performs annually without appearing to give himself the least trouble. . . . There is first the museum of natural history opposite his house, and admirably arranged by himself, then the anatomy museum connected with his dwelling. In the latter is a library disposed in a suite of rooms, each containing works on one subject. There is one where there are all the works on ornithology, in another room all on ichthyology, in another osteology, in another *law* books! etc., etc. . . . The ordinary studio contains no bookshelves. It is a longish room, comfortably fur-

> nished, lighted from above, with eleven desks to stand to, and two low tables, like a public office for so many clerks. But all is for the one man, who multiplies himself as author, and admitting no one into this room, moves as he finds necessary, or as fancy inclines him, from one occupation to another. Each desk is furnished with a complete establishment of inkstand, pens, etc.... There is a separate bell to several desks. The low tables are to sit to when he is tired. The collaborators are not numerous, but always chosen well. They save him every mechanical labour, find references, etc., are rarely admitted to the study, receive orders and speak not.

Cuvier has suffered primarily because posterity has deemed incorrect the two cardinal conclusions that motivated his work in biology and geology—his belief in the fixity of species and his catastrophism. Since being wrong is a primary intellectual sin when we judge the past by its approach to current wisdom, dubious motives must be ascribed to Cuvier. How else can one explain why such a brilliant man went so far astray? Cuvier then becomes an object lesson for aspiring scientists. Cuvier must have failed because he allowed prejudice to cloud objective truth. Conventional theology must have dictated both his creationism and the geological catastrophism that supposedly squeezed our earth into the Mosaic chronology. Consider this assessment of Cuvier presented by a leading modern textbook in geology:

> Cuvier believed that Noah's flood was universal and had prepared the earth for its present inhabitants. The Church was happy to have the support of such an eminent scientist, and there is no doubt that Cuvier's great reputation delayed the acceptance of the more reasonable views that ultimately prevailed.

I devote this essay to defending Cuvier (who ranks, in my judgment, with Darwin and Karl Ernst von Baer as the greatest of nineteenth-century natural historians). But I do not choose to do so in the usual manner of historians—by showing that Cuvier's beliefs were not rooted in irrational prejudice, but both arose from and advanced beyond the social and scientific context of his own time. Nor (obviously) will I defend Cuvier's creationism or more than a sliver of his catastrophism. Instead, I want to argue that Cuvier used the very doctrines for which he stands condemned—creationism and catastrophism—as specific and highly fruitful research strategies for establishing the basis of modern geology—the stratigraphic record of fossils and its attendant long chronology for earth history. Some types of truth may require pursuit on the straight and narrow, but the pathways to scientific insight are as winding and complex as the human mind.

Cuvier is often portrayed as an armchair speculator because his conclusions are now regarded as incorrect and error supposedly arises from aversion to hard data. In fact, he was a committed empiricist. He railed against the prevalent tradition in geology for constructing comprehensive "theories of the earth" with minimal attention to actual rocks and fossils. "Naturalists," he wrote, "seem to have scarcely any idea of the propriety of investigating facts before they construct their systems." (Cuvier correctly includes Hutton, subject of essay 6, among the system builders, although he confesses more sympathy for his Scottish colleague than for most of his ilk.)

Instead, Cuvier argues, we must seek some empirical criterion for unraveling the earth's history. But what shall it be? What has changed with sufficient regularity and magnitude to serve as a marker of time? Cuvier recognized that the lithology of rocks would not do, since limestones and shales look pretty much alike whether they occur at the tops or bottoms of stratigraphic sequences. What about the fossils entombed in rocks?

The idea that fossils reflect history is now so commonplace, we tend to regard it as an ancient truth. It was, however, a contentious issue in Cuvier's day, when debate centered on whether or not species could become extinct—for without extinction, all creatures are coeval and fossils cannot measure time (unless new forms keep accumulating and we can date rocks by first appearances. But a finite earth would seem to preclude continual addition with no subtraction).

Many of Cuvier's illustrious contemporaries (including Thomas Jefferson who, when not preoccupied with other matters, devoted a paper to the subject) argued strongly that extinction could not occur. Cuvier decided that the a priori (and often explicitly biblical) defenses of nonextinction were worthless and that the issue would have to be decided empirically. But previous studies of fossil vertebrates (his specialty) had been undertaken in the mindless manner of mere collection. Fossils had been gathered primarily as curiosities—but scientists must *ask questions* and collect systematically in their light.

> Other naturalists, it is true, have studied the fossil remains of organized bodies; they have collected and represented them by thousands, and their works will certainly serve as a valuable storehouse of materials. But, considering these fossil plants and animals merely in themselves, instead of viewing them in their connection with the theory of the earth; or regarding their petrifactions . . . as mere curiosities, rather than historical documents . . . they have almost always neglected to investigate the general laws affecting their position, or the relation of the extraneous fossils with the strata in which they are found.

Cuvier then provides a two-page compendium of questions, an empiricist's *vade mecum* to combat the older speculative tradition.

> Are there certain animals and plants peculiar to certain strata and not found in others? What are the species that appear first in order, and those which succeed? Do these two kinds of species ever accompany one another? Are there alternations in their appearance; or, in

other words, does the first species appear a second time, and does the second species then disappear?

But this research program for establishing a geological record cannot work unless extinction is a common fact of nature—and ancient creatures are therefore confined to rocks of definite and restricted ages. Cuvier's great four-volume work (*Recherches sur les ossemens fossiles,* "studies on fossil bones") is a long demonstration that fossil bones belong to lost worlds of extinct species.

Cuvier used the comparative anatomy of living vertebrates to assign his fossils to extinct species. Since fossils come in bits and pieces, a tooth here or a femur there, some method must be devised to reconstruct a whole from scrappy parts and to ascertain whether that whole still walks among the living. But what principles shall govern the reconstruction of wholes from parts? Can it be done at all? Cuvier recognized that he must study the anatomy of modern organisms—where we have unambiguous wholes—to learn how to interpret fragments of the past. The second paragraph of his essay presents this program for research:

> As an antiquary of a new order, I have been obliged to learn the art of deciphering and restoring these remains, of discovering and bringing together, in their primitive arrangement, the scattered and mutilated fragments of which they are composed. . . . I had . . . to prepare myself for these enquiries by others of a far more extensive kind, respecting the animals which still exist. Nothing, except an almost complete review of creation in its present state, could give a character of demonstration to the results of my investigations into its ancient state; but that review has afforded me, at the same time, a great body of rules and affinities which are no less satisfactorily demonstrated; and the whole animal kingdom has been subjected to new laws in consequence of this Essay on a small part of the theory of the earth.

As his cardinal rule for reconstruction, Cuvier devised a principle that he called "correlation of parts." Animals are exquisitely designed and integrated structures—perfect Newtonian machines of a sort. Each part implies the next, and a whole lies embodied in the implications of any fragment—a grand version of that immortal commentary on Ezekiel's vision, "the foot bone's connected to the ankle bone. . . ."

Cuvier presents the law of correlation as if it could be applied by reason alone, using the principles of animal mechanics:

> Every organized individual forms an entire system of its own, all the parts of which mutually correspond and concur. . . . Hence none of these separate parts can change their forms without a corresponding change in the other parts of the same animal, and consequently each of these parts, taken separately, indicates all the other parts to which it has belonged. . . . If the viscera of an animal are so organized as only to be fitted for the digestion of recent flesh, it is also requisite that the jaws should be constructed as to fit them for devouring prey; the claws must be constructed for seizing and tearing it to pieces; the teeth for cutting and dividing its flesh; the entire system of the limbs, or organs of motion, for pursuing and overtaking it; and the organs of sense, for discovering it at a distance. . . . Thus, commencing our investigation by a careful survey of any one bone by itself, a person who is sufficiently master of the laws of organic structure, may, as it were, reconstruct the whole animal to which that bone had belonged.

Cuvier's principle of correlation lies behind the popular myth that paleontologists can see an entire dinosaur in a single neck bone. (I believed this legend as a child and once despaired of entering my chosen profession because I could not imagine how I could ever obtain such arcane and wondrous knowledge.) Cuvier's principle may well apply in the most general sense: if I find a jaw with weak peglike teeth, I do not expect to find the sharp claws of a carnivore on the accompanying legs. But a single tooth will not tell me how long the legs were, how sharp the claws, or even how many other teeth the jaw held. Animals are bundles of historical accidents, not perfect and predictable machines.

When a paleontologist does look at a single tooth and says, "Aha, a rhinoceros," he is not calculating through laws of physics, but simply making an empirical association: teeth of this peculiar form (and rhino teeth are distinctive) have never been found in any animal but a rhino. The single tooth implies a horn and a thick hide only because all rhinos share these characters, not because the deductive laws of organic structure declare their necessary connection. Cuvier, in fact, knew perfectly well that he operated by empirical association (and not by logical inference), although he regarded his observational method as an imperfect way station to a future rational morphology:

> As all these relative conformations are constant and regular, we may be assured that they depend upon some sufficient cause; and, since we are not acquainted with that cause, we must here supply the defect of theory by observation, and in this way lay down empirical rules on the subject, which are almost as certain as those deduced from rational principles, especially if established upon careful and repeated observation. Hence, any one who observes merely the print of a cloven hoof, may conclude that it has been left by a ruminant animal, and regard the conclusion as equally certain with any other in physics or in morals.

Since Cuvier didn't know the laws of rational morphology (we now suspect that they do not exist in the form he anticipated), he proceeded by his favorite method of empirical cataloging. He amassed an enormous collection of vertebrate skeletons, and noted an invariant association of parts by repeated observation. He could then use his catalog of recent skeletons to decide whether fossils belong to extinct species. The earth, he argued, has been explored with sufficient care (for large terrestrial mammals at least) that fossil bones outside the range of modern skeletons must represent vanished species.

The four volumes of the 1812 treatise form a single long argument for the fact of extinction, the resultant utility of fossil vertebrates for ascertaining the relative ages of rocks, and the consequent antiquity of the earth. The introductory *Discours préliminaire* sets out basic principles. In the first technical monograph, on mummified remains of the Egyptian ibis, Cuvier finds no difference between modern birds and fossils from the beginning of recorded history as then construed. The present creation therefore has considerable antiquity; if extinct species inhabited still earlier worlds, then the earth must be truly ancient. The next set of monographs discusses the detailed anatomy of large mammals found in the uppermost geological strata—Irish elks, woolly rhinos, and a variety of fossil elephants (mammoths and mastodons). They are similar to modern relatives, but the sizes and shapes of their fragmentary bones lie outside modern ranges and will not correlate with the normal skeletons of living forms (no modern deer could hold up the antlers of an Irish elk). Hence, extinction has occurred and life on earth has a history. The final monographs demonstrate that still older bones belonged to creatures even more unlike modern species. Life's history has a direction—and great antiquity if it has passed through so many cycles of creation and destruction.

Cuvier did not give an evolutionary interpretation to the direction that he discerned, for the very principle that he used to establish extinction—the correlation of parts—precluded evolution in his mind. If an animal's parts are so interdependent that each one implies the exact form of all others, then any change would require a total remodeling of an entire body, and what process can accomplish such a complete and harmonious change all at once? The direction of life's history must reflect a sequence of creations (and subsequent extinctions), each more modern in character.

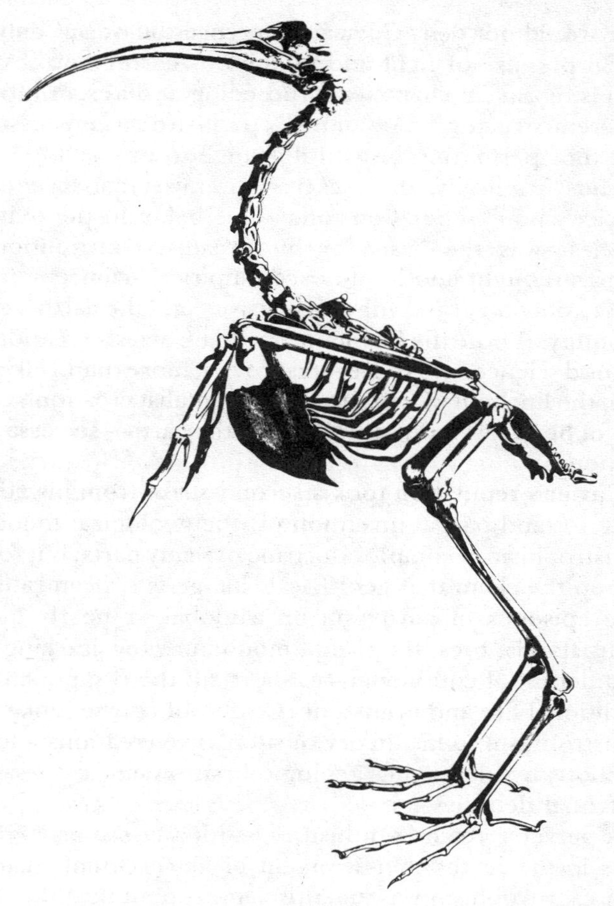

Mummified skeleton of an Egyptian ibis, from Cuvier's *Ossemens fossiles* of 1812. Cuvier showed that this bird is identical with modern ibises and that no organic changes had occurred during the long period from ancient Egypt to today. Since so many changes had occurred in earlier periods, the earth must be ancient.

(We would not deny Cuvier's inference today, but only his initial premise of tight and ubiquitous correlation. Evolution is mosaic in character, proceeding at different rates in different structures. An animal's parts are largely dissociable, thus permitting historical change to proceed.)

Thus, ironically, the incorrect premise that has sealed Cuvier's poor reputation today—his belief in the fixity of species—was the basis for his greatest contribution to human thought and hard-nosed empirical science: a proof that extinction grants life a rich history and the earth a great antiquity. (I note the further irony that Cuvier's creationism—good science in his time—disproved, more than 150 years ago, the linchpin of modern fundamentalist creationism: an age of but a few thousand years for the earth—see essays of section 5).

Cuvier's reputation took a second strike from his adherence to (and partial invention of) the geological theory of catastrophism, a complex doctrine of many parts, but focusing on the claim that geological change is concentrated in rare episodes of paroxysm on a global or nearly global scale: floods, fires, the rise of mountains, the cracking and foundering of continents—in short, all the components of traditional fire and brimstone. Cuvier, of course, linked his catastrophism to his theory of successive creations and extinctions by identifying geological paroxysms as the agent of faunal debacles.

A perverse reading of history had led to the usual claim—as in the textbook assessment of Cuvier cited earlier—that catastrophism was an antiscientific feint by a theological rear guard laboring to place Noah's flood under the aegis of science, and to justify a compression of earth's history into the Mosaic chronology. Of course, if the earth is but a few thousand years old, then we can only account for its vast panoply of observed changes by telescoping them into a few episodes of worldwide destruction. But the converse does not hold: a claim that paroxysms sometimes engulf the earth dictates no conclusion about its age. The earth might be billions of years old, and its changes might still be concentrated in rare episodes of destruction.

Cuvier's eclipse is awash in irony, but no element of his denigration is more curiously unfair than the charge that his catastrophism reflects a theological compromise with his scientific ideals. In the great debates of early-nineteenth-century geology, catastrophists followed the stereotypical method of objective science—empirical literalism. They believed what they saw, interpolated nothing, and read the record of the rocks directly. This record, read literally, is one of discontinuity and abrupt transition: faunas disappear; terrestrial rocks lie under marine rocks with no recorded transitional environments between; horizontal sediments overlie twisted and fractured strata of an earlier age. Uniformitarians, the traditional opponents of catastrophism, did not triumph because they read the record more objectively. Rather, uniformitarians, like Lyell and Darwin, advocated a more subtle and *less* empirical method: use reason and inference to supply the missing information that imperfect evidence cannot record. The literal record is discontinuous, but gradual change lies in the missing transitions. To cite Lyell's thought experiment: if Vesuvius erupted again and buried a modern Italian town directly atop Pompeii, would the abrupt transition from Latin to Italian, or clay tablets to television, record a true historical jump or two thousand years of missing data? I am no partisan of gradual change, but I do support the historical method of Lyell and Darwin. Raw empirical literalism will not adequately map a complex and imperfect world. Still, it seems unjust that catastrophists, who almost followed a caricature of objectivity and fidelity to nature, should be saddled with a charge that they abandoned the real world for their Bibles.

Cuvier's methodology may have been naïve, but one can only admire his trust in nature and his zeal for building a world by direct and patient observation, rather than by fiat or unconstrained feats of imagination. His rejection of received doctrine as a source of necessary truth is, perhaps, most apparent in the very section of the *Discours préliminaire* that might seem, superficially, to tout the Bible as infallible—his defense of Noah's flood. He does argue for a worldwide flood some five thousand years ago, and he does cite the Bible as support. But his thirty-page discussion is a literary and ethnographic compendium of all traditions, from Chaldean to Chinese. And we soon realize that Cuvier has subtly reversed the usual apologetic tradition. He does not invoke geology and non-Christian thought as window dressing for "how do I know, the Bible tells me so." Rather, he uses the Bible as a single source among many of equal merit as he searches for clues to unravel the earth's history. Noah's tale is but one local and highly imperfect rendering of the last major paroxysm.

As a rough rule of thumb, I always look to closing paragraphs as indications of a book's essential character. General treatises in the pontifical mode proclaim a union of all knowledge, or tell us, in no uncertain terms, what *it* all means for man's physical future and moral development. Cuvier's conclusion is revealing in its starkly contrasting style. No drum rolls, no statements about the implications of catastrophism for human history. Cuvier simply presents a ten-page list of outstanding problems in stratigraphic geology. "It appears to me," he writes, "that a consecutive history of such singular deposits would be infinitely more valuable than so many contradictory conjectures respecting the first origin of the world and other planets." He ranges across Europe, up and down the geological column, offering suggestions for empirical work: study recent alluvial deposits of the Po and the Arno, dig in the gypsum quarries of Aix and Paris, collect "gryphites, the cornua ammonis and the entrochi" that may abound in the Black Forest. "We are as yet uninformed of the real position of the stinkstone slate of Oeningen, which is also said to be full of the remains of fresh-water fish."

A man who could end one of the greatest *theoretical* treatises in natural history with a plea for unraveling the stratigraphic position and faunal content of the Oeningen stinkstones knew, in the most profound way, what science is about. We may wallow forever in the thinkable; science traffics in the doable.

8 Agassiz in the Galápagos

I ONCE HAD a gutsy English teacher who used a drugstore paperback called *Word Power Made Easy* instead of the insipid fare officially available. It contained some nifty words, and she would call upon us in turn for definitions. I will never forget the spectacle of five kids in a row denying that they knew what "nymphomania" meant—the single word, one may be confident, that everyone had learned with avidity. Sixth in line was the class innocent; she blushed and then gave a straightforward, accurate definition in her sweet, level voice. Bless her for all of us and our cowardly discomfort; I trust that all has gone well for her since last we met on graduation day.

Nymphomania titillated me to my pubescent core, but two paired words from the same lesson—anachronism and incongruity—interested me more for the eerie feeling they inspired. Nothing elicits a greater mixture of fascination and distress in me than objects or people that seem to be in the wrong time or place. The *little* things that offend a sense of order are the most disturbing. Thus, I was stunned in 1965 to discover that Alexander Kerensky was alive, well, and living as a Russian émigré in New York. Kerensky, the man who preceded the Bolsheviks in 1917? Kerensky, so linked with Lenin and times long past in my thoughts, still among us? (He died, in fact, in 1970, at age 89.)

In July 1981, on a ship headed for the Galápagos Islands, I encountered an incongruity that struck me just as forcefully. I was listening to a lecture when a throwaway line cut right into me. "Louis Agassiz," the man said, "visited the Galápagos and made scientific collections there in 1872." What? The primal creationist, the last great holdout against Darwin, in the Galápagos, the land that stands for evolution and prompted Darwin's own conversion? One might as well let a Christian into Mecca. It seems as incongruous as a president of the United States portraying an inebriated pitcher in the 1926 World Series.

Louis Agassiz was, without doubt, the greatest and most influential naturalist of nineteenth-century America. A Swiss by birth, he was the first great European theorist in biology to make America his home. He had charm, wit, and connections aplenty, and he took the Boston Brahmins by storm. He was an intimate of Emerson, Longfellow, and anyone who really mattered in America's most patrician town. He published and raised money with equal zeal and virtually established natural history as a professional discipline in America; indeed, I am writing this article in the great museum that he built.

But Agassiz's summer of fame and fortune turned into a winter of doubt and befuddlement. He was Darwin's contemporary (two years older), but his mind was indentured to the creationist world view and the idealist philosophy that he had learned from Europe's great scientists. The erudition that had so charmed America's rustics became his undoing; Agassiz could not adjust to Darwin's world. All his students and colleagues became evolutionists. He fretted and struggled, for no one enjoys being an intellectual outcast. Agassiz died in 1873, sad and intellectually isolated but still arguing that the history of life reflects a preordained, divine plan and that species are the created incarnations of ideas in God's mind.

Agassiz did, however, visit the Galápagos a year before he died. My previous ignorance of this incongruity is at least partly excusable, for he never breathed a word about it in any speech or publication. Why this silence, when his last year is full of documents and pronouncements? Why was he there? What impact did those finches and tortoises have upon him? Did the land that so inspired Darwin, fueling his transition from prospective preacher to evolutionary agnostic, do nothing for Agassiz? Is not this silence as curious as the basic fact of Agassiz's visit? These questions bothered me throughout my stay in the Galápagos, but I could not learn the answers until I returned to the library that Agassiz himself had founded more than a century ago.

Agassiz's friend Benjamin Peirce had become superintendent of the Coast Survey. In February of 1871, he wrote to Agassiz offering him the use of the *Hassler,* a steamer fit for deep-sea dredging. I suspect that Peirce had a strong ulterior motive beyond the desire to collect some deep-sea fishes: he hoped that Agassiz's intellectual stagnation might be broken by a long voyage of direct exposure to nature. Agassiz had spent so much time raising money for his museum and politicking for natural history in America that his contact with organisms other than the human kind had virtually ceased. Agassiz's life now belied his famous motto: study nature, not books. Perhaps he could be shaken into modernity by renewed contact with the original source of his fame.

Agassiz understood only too well and readily accepted Peirce's offer. Agassiz's friends rejoiced, for all were saddened by the intellectual hardening of such a great mind. Darwin himself wrote to Agassiz's son: "Pray give my most sincere respects to your father. What a wonderful man he is to think of going round Cape Horn; if he does go, I wish he could go through the Strait of Magellan." The *Hassler* left Boston in December 1871, moved down the eastern coast of South America, fulfilled Darwin's hope by sailing through the Strait of Magellan, passed up the western coast of South America, reached the Galápagos (at the equator, 600 miles off the coast of Ecuador) on June 10, 1872, and finally docked at San Francisco on August 24.

A possible solution to the enigma of Agassiz's silence immediately suggests itself. The Galápagos are pretty much "on the way" along Agassiz's route. Perhaps the *Hassler* only stopped for provisions—just passing by. Perhaps the cruise was so devoted to deep-sea dredging and Agassiz's observations of glaciers in the southern Andes that the

Agassiz, left, and his friend Benjamin Peirce, who arranged for his voyage to the Galápagos. PHOTO COURTESY OF THE GRANGER COLLECTION.

Galápagos provided no special interest or concern.

This easy explanation is clearly incorrect. In fact, Agassiz planned the *Hassler* voyage as a test of evolutionary theory. The dredging itself was not designed merely to collect unknown creatures but to gather evidence that Agassiz hoped would establish the continuing intellectual validity of his lingering creationism. In a remarkable letter to Peirce, written just two days before the *Hassler* set sail, Agassiz stated exactly what he expected to find in the deep dredges.

Agassiz believed that God had ordained a plan for the history of life and then proceeded to create species in the appropriate sequence throughout geological time. God matched environments to the preconceived plan of creation. The fit of life to environment does not record the evolutionary tracking of changing climates by organisms, but rather the construction of environments by God to fit the preconceived plan of creation: "the animal world designed from the beginning has been the motive for the physical changes which our globe has undergone," Agassiz wrote to Peirce. He then applied this curiously inverted argument to the belief, then widespread but now disproved, that the deep oceans formed a domain devoid of change or challenge—a cold, calm, and constant world. God could only have made such an environment for the most primitive creatures of any group. The deep oceans would therefore harbor living representatives of the simple organisms found as fossils in ancient rocks. Since evolution demands progressive change through time, the persistence of these simple and early forms will demonstrate the bankruptcy of Darwinian theory. (I don't think Agassiz ever understood that the principle of natural selection does not predict global and inexorable progress but only adaptation to local environments. The persistence of simple forms in a constant deep sea would have satisfied Darwin's evolutionary theory as well as Agassiz's God. But the depths are not constant, and their life is not primitive.)

The letter to Peirce displays that mixture of psychological distress and intellectual pugnacity so characteristic of Agassiz's opposition to evolution in his later years. He knows that the world will scoff at his preconceptions, but he will pursue them to the point of specific predictions nonetheless—the discovery of "ancient" organisms alive in the deep sea:

> I am desirous to leave in your hands a document which may be very compromising for me, but which I never-

> theless am determined to write in the hope of showing within what limits natural history has advanced toward that point of maturity when science may anticipate the discovery of facts. If there is, as I believe to be the case, a plan according to which the affinities among animals and the order of their succession in time were determined from the beginning . . . if this world of ours is the work of intelligence, and not merely the product of force and matter, the human mind, as a part of the whole, should so chime with it, that, from what is known, it may reach the unknown.

But Agassiz did not sail only to test evolution in the abstract. He chose his route as a challenge to Darwin, for he virtually retraced—and by conscious choice—the primary part of the *Beagle*'s itinerary. The Galápagos were not a convenient way station but a central part of the plot. His later silence becomes more curious.

The *Beagle* did circumnavigate the globe, but Darwin's voyage was basically a surveying expedition of the South American coast. Agassiz's route therefore retraced the essence of Darwin's pathway—physically if not intellectually. One cannot read Elizabeth Agassiz's account of the *Hassler* expedition without recognizing the uncanny (and obviously not accidental) similarity with Darwin's famous account of the *Beagle*'s voyage. (Elizabeth accompanied Louis on the trip.) Darwin concentrated primarily upon geology and so did Agassiz. The trip may have been advertised as a dredging expedition, but Agassiz was most interested in reaching southern South America to test his theory of a global ice age. He had studied glacial striations and moraines in the Northern Hemisphere and had determined that a great ice sheet had once descended from the north. (Striations are scratches on bedrock made by pebbles frozen into the bases of glaciers. Moraines are hills of debris pushed by flowing ice to the fronts and sides of glaciers.) If the ice age had been global, striations and moraines in South America would indicate a spread from Antarctica at the same time. Agassiz's predictions were, in this case, upheld—and he exulted in copious print (faithfully transcribed by Elizabeth and published in the *Atlantic Monthly*).

Darwin was appalled by the rude life and appearance of the "savage" Fuegians and so was Agassiz. Elizabeth recorded their joint impressions: "Nothing could be more coarse and repulsive than their appearance, in which the brutality of the savage was in no way redeemed by physical strength or manliness. . . . They scrambled and snatched fiercely, like wild animals, for whatever they could catch."

If there be any lingering doubt about Agassiz's conscious decision to evaluate Darwin by retracing his experiences, consider this passage, written at sea to his German colleague Carl Gegenbaur:

> I have sailed across the Atlantic Ocean through the Strait of Magellan, and along the western coast of South America to the northern latitudes. Marine animals were, naturally, my primary concern, but I also had a special purpose. I wanted to study the entire Darwinian theory, free from all external influences and former surroundings. Was it not on a similar voyage that Darwin developed his present opinions! I took few books with me . . . primarily Darwin's major works.

I can find few details about Agassiz's stay in the Galápagos. We know that he arrived on June 10, 1872, spent a week or more, and visited five islands, one more than Darwin did. Elizabeth claims that Louis "enjoyed extremely his cruise among these islands of such rare geological and zoological interest." We know that he collected (or rather sat on the rocks while his assistants gathered) the famous iguanas that go swimming in the ocean to eat marine algae (some of his specimens are still in glass jars in our museum). We know that he crossed and greatly admired the bare fields of recently cooled ropy lava "full of the most singular and fantastic details." I walked across a similar field, one that Agassiz could not have seen since it formed during the 1890s. I was mesmerized by the frozen signs of former activity—the undulating, ropy patterns of flow, the burst bubbles, and lengthy cracks of contraction. And I saw Pele's tears, the most beautiful geological object, at small scale, that I have ever witnessed. When highly liquid lava is ejected from small vents, it may emerge as droplets of basalt that build drip castles of iridescent stone about their outlet—tears from Pele, the Hawaiian goddess of volcanoes (not from Martinique's Mount Pelée, which has an extra *e*).

Thus, I return to my original inquiry: if Agassiz went to the Galápagos as a central part of his plan to evaluate evolution by putting himself in Darwin's shoes, what effect did Darwin's most important spot have upon him? In response to this question we have only Agassiz's public silence (and one private communication, to which I will shortly return).

Two nonintellectual reasons may partly explain Agassiz's uncharacteristic reticence. First, despite his productive observations on South American glaciers, the *Hassler* expedition was basically a failure and a profound disappointment—and Agassiz may have chosen largely to forget about it. The dredging equipment never worked properly, and Agassiz recovered no specimens from the deepest oceans. The crew tried its best, but the ship was a misery. Jules Marcou, Agassiz's faithful biographer, wrote: "It was a great, almost a cruel, carelessness to embark a man so distinguished, so old [Agassiz was 64; perhaps concepts of age have changed], and so much an invalid as Agassiz was, in an unseaworthy craft, sailing under the United States flag."

Secondly, Agassiz was ill during much of the voyage, and his listlessness and discomfort increased as he left his beloved southern glaciers and moved into the sultry tropics. (The Galápagos, however, despite their equatorial location, lie in the path of a cool oceanic current and are generally temperate; the northernmost species of penguin inhabits its shores.) Shortly after his return to Harvard, Agassiz wrote to Pedro II, emperor of Brazil (and an old buddy from a previous voyage):

> When I traversed the Strait of Magellan . . . work again became easy for me. The beauty of its sites, the resemblance of the mountains to those of Switzerland, the

> interest that the glaciers awakened in me, the happiness in seeing my predictions affirmed beyond all my hopes—all these conspired to set me on the right course again, even to rejuvenate me. . . . Afterwards, I gradually declined as we advanced towards the tropical regions; the heat exhausted me greatly, and during the month that we spent in Panama I was quite incapable of the least effort.

(For all citations from letters, I have relied upon the originals in Harvard's Houghton Library; none has been published in full before, although several have been excerpted in print. Agassiz wrote with equal facility in French [to Pedro II], German [to Gegenbaur], and English [to Peirce], and I have supplied the translations. I thank my secretary Agnes Pilot for transcribing the Gegenbaur letter into sensible Roman. Agassiz wrote it in the old German script that is all squiggles to me.)

So far as I can tell, Agassiz's only statement about the Galápagos occurs in a private letter to Benjamin Peirce, composed at sea on July 29, 1872, the day after he had written to Gegenbaur (and said nothing about the Galápagos). The letter begins with the lament of all landlubbers: "I fancy this note may reach you in Martha's Vineyard, and I heartily wish I could be there with you, and take some rest from this everlasting rocking." Agassiz continues with his only statement:

> Our visit to the Galapagos has been full of geological and zoological interest. It is most impressive to see an extensive archipelago, of most recent origin, inhabited by creatures so different from any known in other parts of the world. Here we have a positive limit to the length of time that may have been granted for the transformation of these animals, if indeed they are in any way derived from others dwelling in different parts of the world. The Galapagos are so recent that some of the islands are barely covered with the most scanty vegetation, itself peculiar to these islands. Some parts of their

> surface are entirely bare, and a great many of the craters and lava streams are so fresh, that the atmospheric agents have not yet made an impression on them. Their age does not, therefore, go back to earlier geological periods; they belong to our times, geologically speaking; Whence, then, do their inhabitants (animals as well as plants) come? If descended from some other type, belonging to any neighboring land, then it does not require such unspeakably long periods for the transformation of species as the modern advocates of transmutation claim; and the mystery of change, with such marked and characteristic differences between existing species, is only increased, and brought to level with that of creation. If they are autochthones, from what germs did they start into existence? I think that careful observers, in view of these facts, will have to acknowledge that our science is not yet ripe for a fair discussion of the origin of organized beings.

The quotation is long, but it is, so far as I know, exclusive. Its most remarkable aspect is an extreme weakness, almost speciousness, of argument. Agassiz makes but a single point: many animals of the Galápagos live nowhere else. Yet the islands are so young that a slow process of evolution could not have transformed them from related ancestors in the time available. Thus, they were created where we find them (the obvious bottom line, despite Agassiz's final disclaimer that we know too little to reach any firm conclusion).

Two problems: First, although the Galápagos are young (two to five million years for the oldest islands by current reckoning), they are not so pristine as Agassiz indicates. In the letter, Agassiz describes lava flows of the past hundred years or so, and these are virtually devoid of vegetation and so fresh that one can almost see the flow and feel the heat. But Agassiz surely knew that several of the islands (including some on his itinerary) are more densely vegetated and, although not ancient, were surely not formed in a geological yesterday.

Second, Agassiz leaves out the most important aspect of Darwin's argument. The point is not that so many species of the Galápagos are unique but rather that their nearest relatives are invariably found on the adjacent South American mainland. If God created the Galápagos species where we find them, why did he imbue them with signs of South American affinity (especially since the temperate climates and lava habitats of the Galápagos are so different from the tropical environments of the ancestral forms). What sense can such a pattern make unless the species of the Galápagos are modified descendants of South American forms that managed to cross the oceanic barrier? Darwin wrote in the *Voyage of the Beagle:*

> Why, on these small points of land, which within a late geological period must have been covered by the ocean, which are formed by basaltic lava, and therefore differ in geological character from the American continent, and which are placed under a peculiar climate,—why were their aboriginal inhabitats . . . created on American types of organization.

And the famous, poetic statement earlier in the chapter: "We seem to be brought somewhat near to the great fact—that mystery of mysteries—the first appearance of new beings on this earth."

Agassiz could not have misunderstood, for, like Darwin, he was a professional biogeographer. He had also used arguments of geographical distribution as his primary defense for creationism. Why did he skirt Darwin's principal argument? Why did he say so little about the Galápagos and argue so poorly?

I think that we must consider two possibilities as resolutions to the conundrum of Agassiz's silence (or failure to consider the critical points in his one private statement). Perhaps he knew that his argument to Peirce was hokey and inadequate. Perhaps the Galápagos, and the entire *Hassler* voyage, had produced the same change of heart that Darwin had experienced in similar circumstances and Agassiz simply couldn't muster the courage to admit it.

I cannot accept such a resolution. As I said earlier, we see abundant signs of psychological distress and deep sadness in Agassiz's last defenses of creationism. No one enjoys being an intellectual pariah, especially when cast in the role of superannuated fuddy-duddy (the part of ignored but prophetic seer at least elicits moral courage). Yet, however weak his arguments (and they deteriorated as the evidence for evolution accumulated), I sense no failure of Agassiz's resolve. The letter to Peirce seems to represent still another of Agassiz's flawed, but perfectly sincere, defenses of an increasingly indefensible, yet steadfastly maintained, view of life. (Agassiz's last article, posthumously published in the *Atlantic Monthly* in 1874, was a ringing apologia for creationism entitled "Evolution and Permanence of Type.")

I think that we must accept the second resolution: Agassiz said so little about the Galápagos because his visit made preciously little impact upon him. The message is familiar but profound nonetheless. Scientific discovery is not a one-way transfer of information from unambiguous nature to minds that are always open. It is a reciprocal interaction between a multifarious and confusing nature and minds sufficiently receptive (as many are not) to extract a weak but sensible pattern from the prevailing noise. There are no signs on the Galápagos that proclaim: Evolution at work. Open your eyes and ye shall see it. Evolution is an inescapable inference, not a raw datum. Darwin, young, restless, and searching, was receptive to the signal. Agassiz, committed and defensive, was not. Had he not already announced in the first letter to Peirce that he knew what he must find? I do not think he was free to reach Darwin's conclusions, and the Galápagos Islands, therefore, carried no important message for him. Science is a balanced interaction of mind and nature.

Agassiz lived for little more than a year after the *Hassler* docked. James Russell Lowell, traveling abroad, learned of his friend's death from a newspaper and wrote in poetic tribute (quoted from E. Lurie's fine biography of Agassiz,

Louis Agassiz: A Life in Science, University of Chicago Press, 1960):

> . . . with vague, mechanic eyes,
> I scanned the festering news we half despise
> When suddenly,
> As happens if the brain, from overweight
> Of blood, infect the eye,
> Three tiny words grew lurid as I read,
> And reeled commingling: Agassiz is dead!

I do not know. Perhaps a bit of his incorporeal self went up to a higher realm, as some religions assert. Perhaps he saw there old Adam Sedgwick, the great British geologist (and reverend), who at age 87 wrote to Agassiz a year before the *Hassler* sailed:

> It will never be my happiness to see your face again in this world. But let me, as a Christian man, hope that we may meet hereafter in heaven, and see such visions of God's glory in the moral and material universe, as shall reduce to a mere germ everything which has been elaborated by the skill of man.

Be that as it may, Agassiz's ideas had suffered an intellectual death before he ever reached the Galápagos. Life is a series of trades. We have lost the comfort of Agassiz's belief that a superior intelligence directly regulates every step of life's history according to a plan that places us above all other creatures. ("If it had been otherwise," Agassiz wrote to Pedro II in June 1873, "there would be nothing but despair.") We have found a message in the animals and plants of the Galápagos, and all other places, that enables us to appreciate them, not as disconnected bits of wonder, but as integrated products of a satisfactory and general theory of life's history. That, to me at least, is a good trade.

9 | Worm for a Century, and All Seasons

IN THE PREFACE to his last book, an elderly Charles Darwin wrote: "The subject may appear an insignificant one, but we shall see that it possesses some interest; and the maxim 'de minimis lex non curat' [the law is not concerned with trifles] does not apply to science."

Trifles may matter in nature, but they are unconventional subjects for last books. Most eminent graybeards sum up their life's thought and offer a few pompous suggestions for reconstituting the future. Charles Darwin wrote about worms—*The Formation of Vegetable Mould, Through the Action of Worms, With Observations on Their Habits* (1881).

This month* marks the one-hundredth anniversary of Darwin's death—and celebrations are under way throughout the world. Most symposiums and books are taking the usual high road of broad implication—Darwin and modern life, or Darwin and evolutionary thought. For my personal tribute, I shall take an ostensibly minimalist stance and discuss Darwin's "worm book." But I do this to argue that Darwin justly reversed the venerable maxim of his legal colleagues.

Darwin was a crafty man. He liked worms well enough, but his last book, although superficially about nothing else, is (in many ways) a covert summation of the principles of

---

*Darwin died on April 19, 1882 and this column first appeared in *Natural History* in April 1982.

reasoning that he had labored a lifetime to identify and use in the greatest transformation of nature ever wrought by a single man. In analyzing his concern with worms, we may grasp the sources of Darwin's general success.

The book has usually been interpreted as a curiosity, a harmless work of little importance by a great naturalist in his dotage. Some authors have even used it to support a common myth about Darwin that recent scholarship has extinguished. Darwin, his detractors argued, was a man of mediocre ability who became famous by the good fortune of his situation in place and time. His revolution was "in the air" anyway, and Darwin simply had the patience and pertinacity to develop the evident implications. He was, Jacques Barzun once wrote (in perhaps the most inaccurate epitome I have ever read), "a great assembler of facts and a poor joiner of ideas . . . a man who does not belong with the great thinkers."

To argue that Darwin was merely a competent naturalist mired in trivial detail, these detractors pointed out that most of his books are about minutiae or funny little problems—the habits of climbing plants, why flowers of different form are sometimes found on the same plant, how orchids are fertilized by insects, four volumes on the taxonomy of barnacles, and finally, how worms churn the soil. Yet all these books have both a manifest and a deeper or implicit theme—and detractors missed the second (probably because they didn't read the books and drew conclusions from the titles alone). In each case, the deeper subject is evolution itself or a larger research program for analyzing history in a scientific way.

Why is it, we may ask at this centenary of his passing, that Darwin is still so central a figure in scientific thought? Why must we continue to read his books and grasp his vision if we are to be competent natural historians? Why do scientists, despite their notorious unconcern with history, continue to ponder and debate his works? Three arguments might be offered for Darwin's continuing relevance to scientists.

We might honor him first as the man who "discovered"

evolution. Although popular opinion may grant Darwin this status, such an accolade is surely misplaced, for several illustrious predecessors shared his conviction that organisms are linked by ties of physical descent. In nineteenth-century biology, evolution was a common enough heresy.

As a second attempt, we might locate Darwin's primary claim upon continued scientific attention in the extraordinarily broad and radical implications of his proffered evolutionary mechanism—natural selection. Indeed, I have pushed this theme relentlessly in my two previous books, focusing upon three arguments: natural selection as a theory of local adaptation, not inexorable progress; the claim that order in nature arises as a coincidental by-product of struggle among individuals; and the materialistic character of Darwin's theory, particularly his denial of any causal role to spiritual forces, energies, or powers. I do not now abjure this theme, but I have come to realize that it cannot represent the major reason for Darwin's continued *scientific* relevance, though it does account for his impact upon the world at large. For it is too grandiose, and working scientists rarely traffic in such abstract generality.

Everyone appreciates a nifty idea or an abstraction that makes a person sit up, blink hard several times to clear the intellectual cobwebs, and reverse a cherished opinion. But science deals in the workable and soluble, the idea that can be fruitfully embodied in concrete objects suitable for poking, squeezing, manipulating, and extracting. The idea that counts in science must lead to fruitful work, not only to speculation that does not engender empirical test, no matter how much it stretches the mind.

I therefore wish to emphasize a third argument for Darwin's continued importance, and to claim that his greatest achievement lay in establishing principles of *useful* reason for sciences (like evolution) that attempt to reconstruct history. The special problems of historical science (as contrasted, for example, with experimental physics) are many, but one stands out most prominently: Science must identify processes that yield observed results. The results of history lie strewn around us, but we cannot, in principle, directly observe the processes that produced them. How then can we be scientific about the past?

As a general answer, we must develop criteria for inferring the processes we cannot see from results that have been preserved. This is the quintessential problem of evolutionary theory: How do we use the anatomy, physiology, behavior, variation, and geographic distribution of modern organisms, and the fossil remains in our geological record, to infer the pathways of history?

Thus, we come to the covert theme of Darwin's worm book, for it is both a treatise on the habits of earthworms and an exploration of how we can approach history in a scientific way.

Darwin's mentor, the great geologist Charles Lyell, had been obsessed with the same problem. He argued, though not with full justice, that his predecessors had failed to construct a science of geology because they had not developed procedures for inferring an unobservable past from a surrounding present and had therefore indulged in unprovable reverie and speculation. "We see," he wrote in his incomparable prose, "the ancient spirit of speculation revived and a desire manifestly shown to cut, rather than patiently to untie, the Gordian Knot." His solution, an aspect of the complex world view later called uniformitarianism, was to observe the work of present processes and to extrapolate their rates and effects into the past. Here Lyell faced a problem. Many results of the past—the Grand Canyon for example—are extensive and spectacular, but most of what goes on about us every day doesn't amount to much—a bit of erosion here or deposition there. Even a Stromboli or a Vesuvius will cause only local devastation. If modern forces do too little, then we must invoke more cataclysmic processes, now expired or dormant, to explain the past. And we are in catch-22: if past processes were effective and different from present processes, we might explain the past in principle, but we could not be scientific about it because we have no modern analogue in what we can observe. If we rely only upon present processes, we lack sufficient oomph to render the past.

Lyell sought salvation in the great theme of geology: time. He argued that the vast age of our earth provides ample time to render all observed results, however spectacular, by the simple summing of small changes over immense periods. Our failure lay, not with the earth, but with our habits of mind: we had been previously unwilling to recognize how much work the most insignificant processes can accomplish with enough time.

Darwin approached evolution in the same way. The present becomes relevant, and the past therefore becomes scientific, only if we can sum the small effects of present processes to produce observed results. Creationists did not use this principle and therefore failed to understand the relevance of small-scale variation that pervades the biological world (from breeds of dogs to geographical variation in butterflies). Minor variations are the stuff of evolution (not merely a set of accidental excursions around a created ideal type), but we recognize this only when we are prepared to sum small effects through long periods of time.

Darwin recognized that this principle, as a basic mode of reasoning in historical science, must extend beyond evolution. Thus, late in his life, he decided to abstract and exemplify his historical method by applying it to a problem apparently quite different from evolution—a project broad enough to cap an illustrious career. He chose earthworms and the soil. Darwin's refutation of the legal maxim "de minimis lex non curat" was a conscious double-entendre. Worms are both humble and interesting, and a worm's work, when summed over all worms and long periods of time, can shape our landscape and form our soils.

Thus, Darwin wrote at the close of his preface, refuting the opinions of a certain Mr. Fish who denied that worms could account for much "considering their weakness and their size":

> Here we have an instance of that inability to sum up the effects of a continually recurrent cause, which has often retarded the progress of science, as formerly in the case

of geology, and more recently in that of the principle of evolution.

Darwin had chosen well to illustrate his generality. What better than worms: the most ordinary, commonplace, and humble objects of our daily observation and dismissal. If they, working constantly beneath our notice, can form much of our soil and shape our landscape, then what event of magnitude cannot arise from the summation of small effects. Darwin had not abandoned evolution for earthworms; rather, he was using worms to illustrate the general method that had validated evolution as well. Nature's mills, like God's, grind both slowly and exceedingly small.

Darwin made two major claims for worms. First, in shaping the land, their effects are directional. They triturate particles of rock into ever smaller fragments (in passing them through their gut while churning the soil), and they denude the land by loosening and disaggregating the soil as they churn it; gravity and erosive agents then move the soil more easily from high to low ground, thus leveling the landscape. The low, rolling character of topography in areas inhabited by worms is, in large part, a testimony to their slow but persistent work.

Second, in forming and churning the soil, they maintain a steady state amidst constant change. As the primary theme of his book (and the source of its title), Darwin set out to prove that worms form the soil's upper layer, the so-called vegetable mold. He describes it in the opening paragraph:

> The share which worms have taken in the formation of the layer of vegetable mould, which covers the whole surface of the land in every moderately humid country, is the subject of the present volume. This mould is generally of a blackish color and a few inches in thickness. In different districts it differs but little in appearance, although it may rest on various subsoils. The uniform fineness of the particles of which it is composed is one of its chief characteristic features.

Darwin argues that earthworms form vegetable mold by bringing "a large quantity of fine earth" to the surface and depositing it there in the form of castings. (Worms continually pass soil through their intestinal canals, extract anything they can use for food, and "cast" the rest; the rejected material is not feces but primarily soil particles, reduced in average size by trituration and with some organic matter removed.) The castings, originally spiral in form and composed of fine particles, are then disaggregated by wind and water, and spread out to form vegetable mold. "I was thus led to conclude," Darwin writes, "that all the vegetable mould over the whole country has passed many times through, and will again pass many times through, the intestinal canals of worms."

The mold doesn't continually thicken after its formation, for it is compacted by pressure into more solid layers a few inches below the surface. Darwin's theme here is not directional alteration, but continuous change within apparent constancy. Vegetable mold is always the same, yet always changing. Each particle cycles through the system, beginning at the surface in a casting, spreading out, and then working its way down as worms deposit new castings above; but the mold itself is not altered. It may retain the same thickness and character while all its particles cycle. Thus, a system that seems to us stable, perhaps even immutable, is maintained by constant turmoil. We who lack an appreciation of history and have so little feel for the aggregated importance of small but continuous change scarcely realize that the very ground is being swept from beneath our feet; it is alive and constantly churning.

Darwin uses two major types of arguments to convince us that worms form the vegetable mold. He first proves that worms are sufficiently numerous and widely spread in space and depth to do the job. He demonstrates "what a vast number of worms live unseen by us beneath our feet"— some 53,767 per acre (or 356 pounds of worms) in good British soil. He then gathers evidence from informants throughout the world to argue that worms are far more widely distributed, and in a greater range of apparently unfavorable environments, than we usually imagine. He digs to see how deeply they extend into the soil, and cuts one in two at fifty-five inches, although others report worms at eight feet down or more.

With plausibility established, he now seeks direct evidence for constant cycling of vegetable mold at the earth's surface. Considering both sides of the issue, he studies the foundering of objects into the soil as new castings pile up above them, and he collects and weighs the castings themselves to determine the rate of cycling.

Darwin was particularly impressed by the evenness and uniformity of foundering for objects that had once lain together at the surface. He sought fields that, twenty years or more before, had been strewn with objects of substantial size—burned coals, rubble from the demolition of a building, rocks collected from the plowing of a neighboring field. He trenched these fields and found, to his delight, that the objects still formed a clear layer, parallel to the surface but now several inches below it and covered with vegetable mold made entirely of fine particles. "The straightness and regularity of the lines formed by the embedded objects, and their parallelism with the surface of the land, are the most striking features of the case," he wrote. Nothing could beat worms for a slow and meticulous uniformity of action.

Darwin studied the sinking of "Druidical stones" at Stonehenge and the foundering of Roman bathhouses, but he found his most persuasive example at home, in his own field, last plowed in 1841:

> For several years it was clothed with an extremely scant vegetation, and was so thickly covered with small and large flints (some of them half as large as a child's head) that the field was always called by my sons "the stony field." When they ran down the slope the stones clattered together. I remember doubting whether I should live to see these larger flints covered with vegetable mould and turf. But the smaller stones disappeared before many years had elapsed, as did every one of the

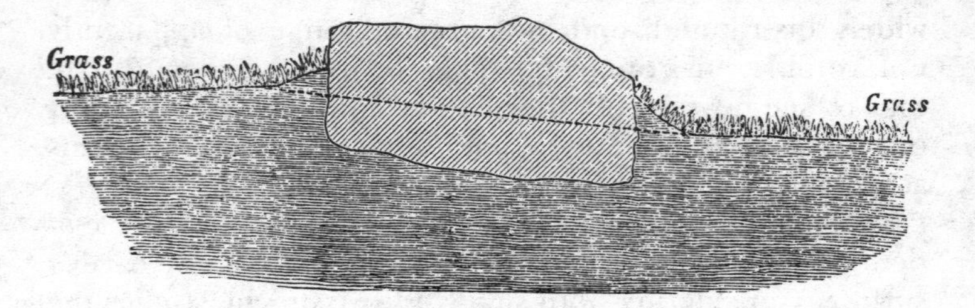

Section through one of the fallen Druidical stones at Stonehenge, showing how much it had sunk into the ground. Scale ½ inch to 1 foot.

An original illustration from Darwin's worm book showing the foundering of large stones by the action of worms.

> larger ones after a time; so that after thirty years (1871) a horse could gallop over the compact turf from one end of the field to the other, and not strike a single stone with his shoes. To anyone who remembered the appearance of the field in 1842, the transformation was wonderful. This was certainly the work of the worms.

In 1871, he cut a trench in his field and found 2.5 inches of vegetable mold, entirely free from flints: "Beneath this lay coarse clayey earth full of flints, like that in any of the neighboring ploughed fields. . . . The average rate of accumulation of the mould during the whole thirty years was only .083 inch per year (i.e., nearly one inch in twelve years)."

In various attempts to collect and weigh castings directly, Darwin estimated from 7.6 to 18.1 tons per acre per year. Spread out evenly upon the surface, he calculated that from 0.8 to 2.2 inches of mold would form anew every ten years. In gathering these figures, Darwin relied upon that great, unsung, and so characteristically British institution—the corps of zealous amateurs in natural history, ready to endure any privation for a precious fact. I was particularly impressed by one anonymous contributor: "A lady," Darwin tells us, "on whose accuracy I can implicitly rely, offered to collect during a year all the castings thrown up on two separate square yards, near Leith Hill Place, in Surrey." Was she the analogue of a modern Park Avenue woman of means, carefully scraping up after her dog: one bag for a cleaner New York, the other for Science with a capital S?

The pleasure of reading Darwin's worm book lies not only in recognizing its larger point but also in the charm of detail that Darwin provides about worms themselves. I would rather peruse 300 pages of Darwin on worms than slog through 30 pages of eternal verities explicitly preached by many writers. The worm book is a labor of love and intimate, meticulous detail. In the book's other major section, Darwin spends 100 pages describing experiments to determine which ends of leaves (and triangular paper cutouts, or abstract "leaves") worms pull into their burrows first. Here we also find an overt and an underlying theme, in this case leaves and burrows versus the evolution of instinct and intelligence, Darwin's concern with establishing a usable definition of intelligence, and his discovery (under that definition) that intelligence pervades "lower" animals as well. All great science is a fruitful marriage of detail and generality, exultation and explanation. Both Darwin and his beloved worms left no stone unturned.

I have argued that Darwin's last book is a work on two levels—an explicit treatise on worms and the soil and a covert discussion of how to learn about the past by studying the present. But was Darwin consciously concerned with establishing a methodology for historical science, as I have argued, or did he merely stumble into such generality in his last book? I believe that his worm book follows the pattern of all his other works, from first to last: every compendium on minutiae is also a treatise on historical reasoning—and each book elucidates a different principle.

Consider his first book on a specific subject, *The Structure and Distribution of Coral-Reefs* (1842). In it, he proposed a theory for the formation of atolls, "those singular rings of coral-land which rise abruptly out of the unfathomable ocean," that won universal acceptance after a century of

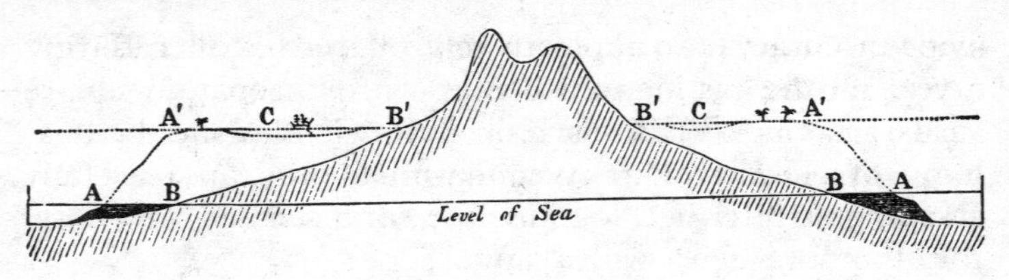

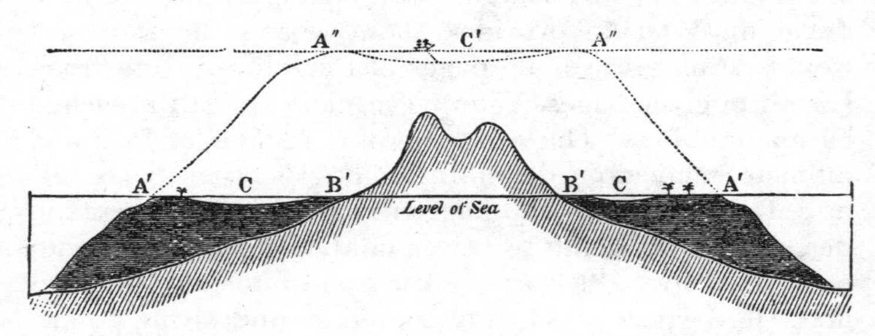

Darwin's original illustration for his theory of coral reefs. Top figure: lower solid line, stage 1, a fringing reef (AB) abuts the shore line. Island sinks (level of sea rises) to upper dotted line, stage 2, barrier reef (A′) separated from sinking island by a lagoon (C).
Bottom figure: lower solid line, stage 2, barrier reef (copied from upper dotted line of top figure). Island sinks further (below level of sea) to upper dotted line, stage 3, an atoll (A″), enlarged lagoon (C′) marks previous location of sunken island.

subsequent debate. He argued that coral reefs should be classified into three categories—fringing reefs that abut an island or continent, barrier reefs separated from island or continent by a lagoon, and atolls, or rings of reefs, with no platform in sight. He linked all three categories with his "subsidence theory," rendering them as three stages of a single process: the subsidence of an island or continental platform beneath the waves as living coral continues to grow upward. Initially, reefs grow right next to the platform (fringing reefs). As the platform sinks, reefs grow up and outward, leaving a separation between sinking platform and living coral (a barrier reef). Finally the platform sinks entirely, and a ring of coral expresses its former shape (an atoll). Darwin found the forms of modern reefs "inexplicable, excepting on the theory that their rocky bases slowly and successively sank beneath the level of the sea, whilst the corals continued to grow upwards."

This book is about coral, but it is also about historical reasoning. Vegetable mold formed fast enough to measure its rate directly; we capture the past by summing effects of small and observable present causes. But what if rates are too slow, or scales too large, to render history by direct observation of present processes? For such cases, we must develop a different method. Since large-scale processes begin at different times and proceed at diverse rates, the varied stages of different examples should exist simultaneously in the present. To establish history in such cases, we must construct a theory that will explain a series of present phenomena as stages of a single historical process. The method is quite general. Darwin used it to explain the formation of coral reefs. We invoke it today to infer the history of stars. Darwin also employed it to establish organic evolution itself. Some species are just beginning to split from their ancestors, others are midway through the process, still others are on the verge of completing it.

But what if evidence is limited to the static object itself? What if we can neither watch part of its formation nor find several stages of the process that produced it? How can we infer history from a lion? Darwin treated this problem in his treatise on the fertilization of orchids by insects (1862); the book that directly followed the *Origin of Species.* I have discussed his solution in several essays (1, 4, 11 and *The Panda's Thumb*) and will not dwell on it here: we infer history from imperfections that record constraints of descent. The "various contrivances" that orchids use to attract insects and attach pollen to them are the highly altered parts of ordinary flowers, evolved in ancestors for other purposes. Orchids work well enough, but they are jury-rigged to succeed because flowers are not optimally constructed for modification to these altered roles. If God wanted to make insect attractors and pollen stickers from scratch, he would certainly have built differently.

Thus, we have three principles for increasing adequacy of data: if you must work with a single object, look for imperfections that record historical descent; if several objects are available, try to render them as stages of a single historical process; if processes can be directly observed, sum up their effects through time. One may discuss these principles directly or recognize the "little problems" that Darwin used to exemplify them: orchids, coral reefs, and worms—the middle book, the first, and the last.

Darwin was not a conscious philosopher. He did not, like Huxley and Lyell, write explicit treatises on methodology. Yet I do not think he was unaware of what he was doing, as he cleverly composed a series of books at two levels, thus expressing his love for nature in the small and his ardent desire to establish both evolution and the principles of historical science. I was musing on this issue as I completed the worm book two weeks ago. Was Darwin really conscious of what he had done as he wrote his last professional lines, or did he proceed intuitively, as men of his genius sometimes do? Then I came to the very last paragraph, and I shook with the joy of insight. Clever old man; he knew full well. In his last words, he looked back to his beginning, compared those worms with his first corals, and completed his life's work in both the large and the small:

> The plough is one of the most ancient and most valuable of man's inventions; but long before he existed the land was in fact regularly ploughed, and still continues to be thus ploughed by earthworms. It may be doubted whether there are many other animals which have played so important a part in the history of the world, as have these lowly organized creatures. Some other animals, however, still more lowly organized, namely corals, have done more conspicuous work in having constructed innumerable reefs and islands in the great

oceans; but these are almost confined to the tropical zones.

At the risk of unwarranted ghoulishness, I cannot suppress a final irony. A year after publishing his worm book, Darwin died on April 19, 1882. He wished to be buried in the soil of his adopted village, where he would have made a final and corporeal gift to his beloved worms. But the sentiments (and politicking) of fellow scientists and men of learning secured a guarded place for his body within the well-mortared floor of Westminster Abbey. Ultimately the worms will not be cheated, for there is no permanence in history, even for cathedrals. But ideas and methods have all the immortality of reason itself. Darwin has been gone for a century, yet he is with us whenever we choose to think about time.

10 | A Hearing for Vavilov

IN 1936, TROFIM D. LYSENKO, struggling to reform Russian agricultural science on discredited Lamarckian principles, wrote: "I am not fond of controversy in matters concerning theory. I am an ardent controversialist only when I see that in order to carry out certain practical tasks I must remove the obstacles that stand in the way of my scientific activities."

As his practical task, Lysenko set out to "alter the nature of plants in the direction we desire by suitable training." He argued that previous failure to produce rapid and heritable improvements in important crop plants must be laid to the bankrupt ideology of bourgeois science, with its emphasis on sterile academic theory and its belief in Mendelian genes, which do not respond directly to the prodding of breeders but change only by accidental and random mutation. The criterion of a more adequate science must be success in improved breeding.

"The better we understand the laws of development of plant and animal forms," he wrote, "the more easily and quickly will we be able to create the forms we need in accordance with our wishes and plans." What "laws of development" could be more promising than the Lamarckian claim that altered environments can directly induce heritable changes in desired directions? If only Nature worked this way! But she does not, and all Lysenko's falsified data and vicious polemics budged her not one inch.

If Lysenko's "obstacles" had been disembodied ideas alone, the history of Russian genetics might have been spared some of its particular tragedy. But ideas emanate from people, and the obstacles designated for removal were necessarily human. Nikolai Ivanovich Vavilov, Russia's leading Mendelian geneticist and director of the All-Union Lenin Academy of Agricultural Sciences centered in Leningrad, served as a focal point for Lysenko's attacks in 1936. Lysenko castigated Vavilov for his general Mendelian views, but any geneticist might have served equally well for such generalized target practice. Lysenko singled out Vavilov for a more specific and personal theory (and the subject of this column)—the so-called law of homologous series in variation.

Twelve years later, following the devastation of war, Lysenko had triumphed. His infamous address, "The Situation in Biological Science," read at the 1948 session of the Lenin Academy of Agricultural Sciences, contains as the first statement of its summary what may well be the most chilling passage in all the literature of twentieth-century science.

> The question is asked in one of the notes handed to me, "What is the attitude of the Central Committee of the Party to my report?" I answer: The Central Committee of the Party has examined my report and approved it [*Stormy applause. Ovation. All rise*].

Following another ten pages of rhetoric and invective, Lysenko concludes: "Glory to the great friend and protagonist of science, our leader and teacher, Comrade Stalin! [*All rise. Prolonged applause.*]"

Nikolai Vavilov was unable to attend the 1948 meeting. He had been arrested in August 1940 while on a collecting expedition in the Ukraine. In July 1941, he was sentenced to death for agricultural sabotage, spying for England, maintaining links with émigrés, and belonging to a rightist organization. The sentence was commuted to ten years imprisonment, and Vavilov was moved to the inner prison of the NKVD in Moscow. In October, he was evacuated to the Saratov prison where he spent several months in an underground death cell, suffering from malnutrition. He died, still a prisoner, in January 1943.

What is Vavilov's "law of homologous series in variation," and how did it provide Lysenko with rhetorical leverage? Vavilov published this law, the guiding principle for much of his practical work in agricultural genetics, in 1920 and revised it in 1935. It was printed in English in the prestigious *Journal of Genetics* in 1922 (vol. 12, pp. 48–89).

Vavilov was perhaps the world's leading expert on the biogeography of wheat and other cereals. He traveled throughout the world (thereby leaving himself vulnerable to trumped-up charges of espionage), collecting varieties of plants from their natural habitats and establishing the world's largest "bank" of genetic variation within major agricultural species. As he collected natural races of wheat, barley, oats, and millet over a large range of environments and places, he noticed that strikingly similar series of varieties could be found within the different species of a genus and often within species of related genera as well.

He collected, for example, a large number of geographical races within the species of common wheat, *Triticum vulgare.* These varied in complex sets of traits, including color of the ears and seeds, form of the ears (bearded or beardless, smooth or hairy), and season of maturation. Vavilov was then surprised and delighted to find virtually the same combinations of characters in varieties of two closely related species, *T. compactum* and *T. spelta.*

He then studied rye *(Secale cereale),* a species in a genus closely related to wheat but previously regarded as much more limited in its geographical variation. Yet, as Vavilov and his assistants collected rye throughout European and Asiatic Russia, Iran, and Afghanistan, he found not only that its differentiation matched wheat in extent but also that its races displayed the same sets of characters, with the same variations in color, form, and timing of growth.

The similarities in series of races were so precise and complete between related species that Vavilov felt he could predict the existence of undiscovered varieties within one species after finding their parallel forms in another species. In 1916, for example, he found several varieties of wheat without ligules in Afghanistan (ligules are thin membranes that grow from the base of the leaf blade and surround the stem in many grasses). This discovery suggested that varieties of rye without ligules should also exist, and he grew them from seeds collected in Pamir in 1918. He predicted that durum wheat *(Triticum durum),* then represented exclusively by spring varieties, should also have winter forms since related species do—and he found them in 1918 in an isolated region of northern Iran.

Vavilov's observation would have engendered less controversy had he not interpreted it, indeed overinterpreted it, in a manner uncongenial with strictly Darwinian or Lamarckian views. He might have argued that his series of parallel varieties represented similar adaptations of different genetic systems to common environments that engendered natural selection in the same direction. Such an interpretation would have satisfied Darwinian preferences for random variation, with evolutionary change imposed by natural selection. (It could also have been distorted by Lysenko into a claim that environments directly altered the heredity of plants in favorable ways.)

But Vavilov proposed a different explanation more in tune with non-Darwinian (though not anti-Darwinian) themes still popular during the 1920s: he claimed that the parallel series of varieties represented identical responses of the same genetic systems, inherited in toto from species to related species. Thus, in evolutionary parlance, his series were "homologous"—hence the name of his law. (Homologies are similarities based on inheritance of the same genes or structures from a common ancestor. Similarities forged within different genetic systems by selective pressures of similar environments are called analogies.)

Vavilov argued that new species arise by developing genetic differences that preclude interbreeding with related species. But the new species is not genetically distinct from its ancestor in all ways. Most of the ancestor's genetic system remains intact; only a limited number of genes are altered. The parallel varieties, then, represent a "playing out" of the same genetic capacities inherited as blocks from species to related species.

Such an interpretation is not anti-Darwinian because it does not deny an important role to natural selection. While each variety may represent a predictable latent capacity, its expression in any climate or geographical region still requires selection to preserve the adaptive variant and to eliminate others. But such an explanation does conflict with the spirit of strict Darwinism because it weakens or compromises the cardinal tenet that selection is *the* creative force in evolution. Random, or undirected, variation plays a crucial role in the Darwinian system because it establishes the centrality of selection by guaranteeing that evolutionary *change* cannot be ascribed to variation itself. Variation is only raw material. It arises in all directions or, at least, is not preferentially ordered in adaptive ways. Hence, direction is imposed by natural selection, slowly preserving and accumulating, generation after generation, the variations that render organisms better adapted to local environments.

But what if variation is not fortuitous and undirected but strongly channeled along certain paths? Then only a limited number of changes are possible, and they record the "internal" constraints of inheritance as much as the action of selection. Selection is not dormant; it still determines which of several possibilities reaches expression in any one climate or area. But if possibilities are strictly limited, and if a species displays all of them among its several varieties, then this range of form cannot be ascribed only to selection acting upon fortuitous variation.

Moreover, this explanation for new varieties compromises the cardinal principle of creativity for natural selection. The variations are predictable results within their genetic system. Their occurrence is almost foreordained. The role of natural selection is negative. It is an executioner only. It eliminates the variants unfit in any given environment, thus preserving the favored form that had to arise eventually.

Vavilov interpreted his law of homologous series in this non-Darwinian manner. "Variation," he wrote, "does not take place in all directions, by chance and without order, but in distinct systems and classes analogous to those of crystallography and chemistry. The same great divisions [of organisms] into orders and classes manifest regularities and repetitions of systems." He cites the case of "several varieties of vetches so similar to ordinary lentils in the shape, color, and size of their seeds, that they cannot be separated by any sorting machine." He agrees that the extreme similarity in any one spot is a product of selection—unconscious selection in agricultural sorting machines. But the agent of selection was, literally in this case, only a sieve that preserved one variant among many. The proper variant existed already as a realized product of an inherited set of possibilities.

> The role of natural selection in this case is quite clear. Man unconsciously, year after year, by his sorting machines separated varieties of vetches similar to lentils in size and form of seeds, and ripening simultaneously with lentils. The same varieties certainly existed long before selection itself, and the appearance of their series, irrespective of any selection, was in accordance with the laws of variation.

Vavilov, overenthused with his own idea, went on. He became intoxicated with the notion that his law might represent a principle of ordering that would render biology as exact and experimental as the "hard" sciences of physics and chemistry. Perhaps genetic systems are composed of "elements." Perhaps the geographical varieties of species are predictable "compounds" that arise inevitably from the union of these elements in specified mixtures. If so, the ranges of biological form within a species might be laid out in a table of possibilities similar to the periodic table of chemical elements. Evolution might be deduced from genetic structure itself; environment can only act to preserve inherent possibilities.

He advocated an explicit "analogy with chemistry" in the concluding section of his 1922 paper and wrote: "New forms have to fill vacancies in a system." He experimented with a style of notation that expressed varieties of a species as a chemical formula and advocated "the analogy of homological series of plants and animals with systems and classes of crystallography with definite chemical structures." One zealous supporter commented that "biology has found its Mendeleev."

Vavilov moderated his views during the 1920s and early 1930s. He learned that some of the parallel varieties between species are not based upon the same genes after all, but represent the similar action of selection upon different sources of variation. In these cases, the varieties are analogous, not homologous, and the Darwinian explanation must be preferred. He wrote in 1937:

> We underestimated the variability of the genes themselves. . . . At the time we thought that the genes possessed by close species were identical; now we know that this is far from the case, that even very closely related species which have externally similar traits are characterized by many different genes. By concentrating our attention on the variability itself, we gave insufficient attention to the role of selection.

Still, Vavilov continued to champion the importance and validity of his law, and he continued to advocate the analogy with chemistry in only slightly weakened form.

Unfortunately, in the deepest sense, Vavilov had left himself open to Lysenko's polemical attack. The law of homologous series provided Lysenko with important ammunition, and Vavilov's overextended chemical analogy deepened his troubles. Lysenko caricatured Vavilov's law in 1936 by presenting ridiculous examples involving species too distantly related to present parallel series in Vavilov's system: "In nature we find apple trees with round fruit, hence there must or can be trees with round pears, cherries, grapes, etc."

Lysenko's ideological attack was more vicious. He made two major charges involving both parts of that catchword for official Soviet philosophy—dialectical materialism. Vavilov's law, he claimed, was undialectical because it located the source of organic change within the genetic systems of organisms themselves and not in the interaction (or dialectic) between organism and environment. Secondly, Lysenko charged that the law of homologous series was "idealist" rather than materialist because it viewed the evolutionary history of a species as prefigured in the unrealized (and therefore nonmaterial) capacity of an inherited genetic system.

Evolution, Lysenko charged, is almost an illusion in Vavilov's scheme. It represents a mere playing out of inherited potentials, not the development of anything new. It expresses the bourgeois penchant for stability by depicting apparent change as a superficial expression of underlying constancy. According to Vavilov's law, Lysenko charged,

> New forms result not from the development of old forms, but from a reshuffling, a recombination of already existing hereditary corpuscles. . . . All the existing species existed in the past, only in less diverse forms; but every form was richer in potentialities, in its collection of genes.

Madness often displays a perverse but cogent reason in its own terms; and we must admit that Lysenko did identify and exploit the true weaknesses in Vavilov's argument. Vavilov did underplay the creative role of environment, and his chemical analogy did betray a belief in prefigured potentiality as the source of later, and in some sense illusory, change. But Lysenko, who was also both a charlatan and a cruel polemicist, was equally undialectical (despite his protestations to the contrary) in viewing plants as putty before a molding environment.

Vavilov died in the name of a phony Lamarckism. He became a legitimate martyr in the West, but his ideas did not flourish as a result. The law of homologous series, the organizing theme of his evolutionary work, was ignored in the name of an overly strict Darwinism. Vavilov's law did not directly contradict Darwinian principles, but its emphasis on constraints of inheritance and channeled variation fit poorly with the favored Darwinian theme of random variation and the guidance of evolutionary change by natural selection. It was therefore neglected and relegated to the shelf of antiquated theories that had implicated variation itself as a directing force in evolution. I have consulted all the founding documents of the "modern synthesis," the movement that established our present version of Darwinism between the late 1930s and the 1950s. Only two mention Vavilov's law of homologous series, each in less than one paragraph.

Yet I feel that in his imperfect way Vavilov had glimpsed something important. In more modern terms, new species do not inherit an adult form from their ancestors. They receive a complex genetic system and a set of developmental pathways for translating genetic products through embryology and later growth into adult organisms. These pathways do constrain the expression of genetic variation; they do channel it along certain lines. Natural selection may choose any spot along the line, but it may not be able to move a species off the line—for selection can only act upon the variation presented to it. In this sense, constraints of variation may direct the paths of evolutionary change as much as selection acting in its Darwinian role as a creative force.

I have found Vavilov's views very helpful in reorienting my own thinking in directions I regard as more fruitful than my previous unquestioned conviction that selection manufactures almost every evolutionary change. In studying the relationship of brain size to body size, biologists find that brains increase only one-fifth to two-fifths as fast as bodies in comparisons of closely related mammals differing only (or primarily) in body size—adults within a single species, breeds of domestic dogs, chimpanzees versus gorillas, for example. For ninety years, the large literature has centered on speculations about the adaptive reasons for this relationship, based upon the (usually unstated) assumption that it must arise as the direct product of natural selection.

But my colleague Russell Lande recently called my attention to several experiments on mice selected over several generations for larger body size alone. As these mice increased in size across generations, their brains enlarged at the characteristic rate—a bit more than one-fifth as fast as body size. Since we know that these experiments included no selection upon brain size, the one-fifth rate must be a side product of selected increase in body size alone. Since the one-fifth to two-fifths rate appears again and again in diverse lineages of mammals, and since it may record a nonadaptive response of brains to selection for larger bodies within mammalian developmental systems, the parallel sets of races and species arrayed along the one-fifth to two-fifths slope in carnivores, rodents, ungulates, and primates are non-Darwinian homologous series in Vavilov's sense.

In personal research on the West Indian land snail *Cerion,* my colleague David Woodruff and I find the same two morphologies again and again in all the northern islands of the Bahamas. Ribby, white or solid-colored, thick, and roughly rectangular shells inhabit rocky coasts at the edges of banks where islands drop abruptly into deep seas. Smooth, mottled, thinner, and barrel-shaped shells inhabit calmer and lower coasts at the interior edges of banks, where islands cede to miles of shallow water. The easiest, and usual, conclusion would view ribby shells on all islands as closely related and smooth shells as members of another coherent group. But we believe that the complex set of characters forming the ribby and smooth morphologies arise independently, again and again. On the islands of Little Bahama Bank, both ribby and smooth animals share a distinctive genital anatomy. On the islands of Great Bahama Bank, both ribby and smooth develop an equally distinctive, but different, kind of penis. The ecology of rocky versus calm coasts may select for ribby and smooth morphologies as adaptations, but the coordinated appearance of the half dozen distinctive traits of each morphology may represent a channeling of available variation to produce homologous series (ribby and smooth varieties) in different lineages (defined by genital anatomy).

A complete theory of evolution must acknowledge a balance between "external" forces of environment imposing selection for local adaptation and "internal" forces representing constraints of inheritance and development. Vavilov placed too much emphasis on internal constraints and downgraded the power of selection. But Western Darwinians have erred equally in practically ignoring (while acknowledging in theory) the limits placed upon selection by structure and development—what Vavilov and the older biologists would have called "laws of form." We need, in short, a real dialectic between the external and internal factors of evolution.

Vavilov's personal tragedy cannot be undone. But he has been rehabilitated in Russia, where the All-Union Society of Geneticists and Selectionists now bears his name. We who view him as a martyr and champion his name while ignoring his ideas would do well to reconsider the older non-Darwinian tradition that he represented. Combined with our legitimate conviction about the power of selection, the principle of homologous series (and other "laws of form") might foster an evolutionary theory truly synthetic in its integration of development and organic form with a body of principles now dominated by ecology and the effects of selection upon single genes and traits.

3 Adaptation and Development

11 Hyena Myths and Realities

I FREELY ADMIT that the spotted, or laughing, hyena is not the loveliest animal to behold. Still, it scarcely deserved the poor reputation imposed upon it by our illustrious forebears. Three myths about hyenas helped to inspire the loathing commentary of ancient texts.

Hyenas, first of all, were regarded as scavengers and consumers of carrion. In his *Natural History,* Pliny the Elder (A.D. 23–79) spoke of them as the only animals that dig up graves in search of corpses *(ab uno animali sepulchra erui inquisitione corporum).* Conrad Gesner, the great sixteenth-century cataloger of natural history, reported that they gorge themselves so gluttonously after finding a corpse that their bellies swell to become taut as a drum. They then seek a narrow place between two trees or stones, force themselves through it, and extrude the remains of their meal simultaneously at both ends.

Hans Kruuk, who spent years studying spotted hyenas on their home turf (the plains of East Africa), has labored to dispel these ancient myths (see his book *The Spotted Hyena,* University of Chicago Press, 1972). He reports that hyenas will scavenge when they get the opportunity. (Almost all carnivores, including the noble lion, will happily feast upon the dead product of another animal's labor.) But spotted hyenas live in hunting clans of up to eighty animals. Each clan controls a territory and kills most of its own food—

mainly zebra and wildebeest—in communal, nocturnal pursuit.

As a second insult, hyenas were widely regarded as hybrids. Sir Walter Raleigh excluded them from Noah's ark since he believed that God had only saved thoroughbreds. Hyenas were reconstituted after the flood through the unnatural union of a dog and cat. In fact, the three living species of hyena form a family of their own within the order Carnivora. They are most closely related to the mustelids (weasels and their allies).

As a final, phony blot on their escutcheon, and in the unkindest cut of all, many ancient writers charged that hyenas were hermaphrodites, bearing both male and female organs. Medieval bestiaries, always trying to draw a moral lesson from the depravity of beasts, focused on this supposed sexual ambivalence. A twelfth-century document, translated by T. H. White, declared:

> Since they are neither male nor female, they are neither faithful nor pagan, but are obviously the people concerning whom Solomon said: "A man of double mind is inconstant in all his ways." About whom also the Lord said: "Thou canst not serve God and Mammon."

But hyenas also had some formidable defenders against this particular calumny. Aristotle himself had declared in the *Historia animalium:* "The statement is made that the hyena has both male and female sexual organs; but this is untrue."

Aristotle—and not for the first time—was right of course. But the legend had arisen for a good reason. Female hyenas are virtually indistinguishable from males. Their clitoris is enlarged and extended to form an organ of the same size, shape, and position as the male penis. It can also be erected. Their labia have folded up and fused to form a false scrotum that is not discernibly different in external form or location from the true scrotum of males. It even contains fatty tissue forming two swellings easily mistaken for testicles. Authors of the most recent paper on spotted hyenas found the appearance of males and females "so close that sex could only be determined with certainty by palpation of the scrotum. Testes could be located in the scrotum of the male compared with soft adipose tissue in the false scrotum of the female."

British zoologist L. Harrison Matthews wrote the most extensive anatomical description of the hyena's sexual anatomy in 1939. He described the peniform clitoris, emphasizing that it is no smaller than the male penis, is equally constricted to a single slitlike opening at the tip, and is as subject to erection as its male counterpart. He concluded his dry and precise pages of description with as forceful a statement of wonder as measured British scientific prose would allow: "It is probably one of the most unusual of the forms which the external orifice of the urogenital canal takes amongst female mammals."

Harrison Matthews also investigated the interesting question of how hyenas do it, given a female orifice no larger than the slit of a male's penis. "In the pre-pubertal state," he writes, "these functions are obviously impossible, owing to the minute size of the opening." But as the female matures the slit gradually lengthens and "creeps down round the ventral surface . . . travelling down the midline" until it forms an orifice 1.5 cm long and extending from the tip of the clitoris to its base. This lengthening of the slit and a subsequent enlargement of the nipples following pregnancy and parturition help distinguish older females from males. We can now understand the basis for ancient myths that hyenas were either simultaneous hermaphrodites (bearing male and female organs at the same time) or male for part of their life and then female.

Nature's oddities cry out for explanation, and we therefore ask what advantages females gain from looking like males. Immediately, we come upon the other most striking oddity of hyena biology: females not only resemble males, they are also larger than males, contrary to the usual pattern in mammals, including humans (but see essay 1 for a discussion of the reverse pattern in most other animals). Females in Kruuk's East African clans averaged 120 pounds in body

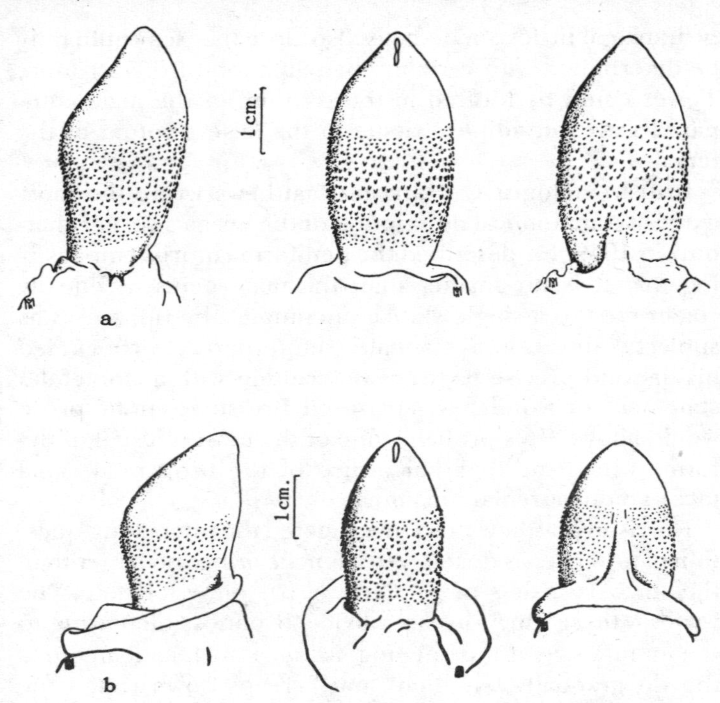

Similarity of male and female genitalia in the spotted hyena. Top row, views of the male penis. Bottom row, similar views of the female clitoris. FROM HARRISON MATTHEWS, 1939.

weight versus 107 pounds for males. Moreover, they lead the clans in hunting and defense of territory and are generally dominant over males in individual contacts. Dominance is not merely a result of larger size because females also rank higher than larger males if the discrepancy in size is not too great.

Although the female hyena's assumption of what are usually male roles in mammals is probably related to its evolution of sexual structures that mimic male organs, the link between these phenomena is not immediately clear. It cannot have much to do with sexual performance itself for, if anything, the female "penis" is a hindrance to copulation until its opening enlarges and its form departs from that of the male.

Kruuk suggests that the strong mimicry arose in connection with a common behavior in hyenas called the "meeting ceremony." Hyenas live in clans that defend territories and engage in communal hunting. But individuals also spend much of their time as solitary wanderers searching the landscape for carrion. To maintain cohesion in clans and to keep strangers away, hyenas must develop a mechanism for recognizing each other and reintegrating solitary wanderers into their proper clan.

When two hyenas of the same clan meet, they stand side

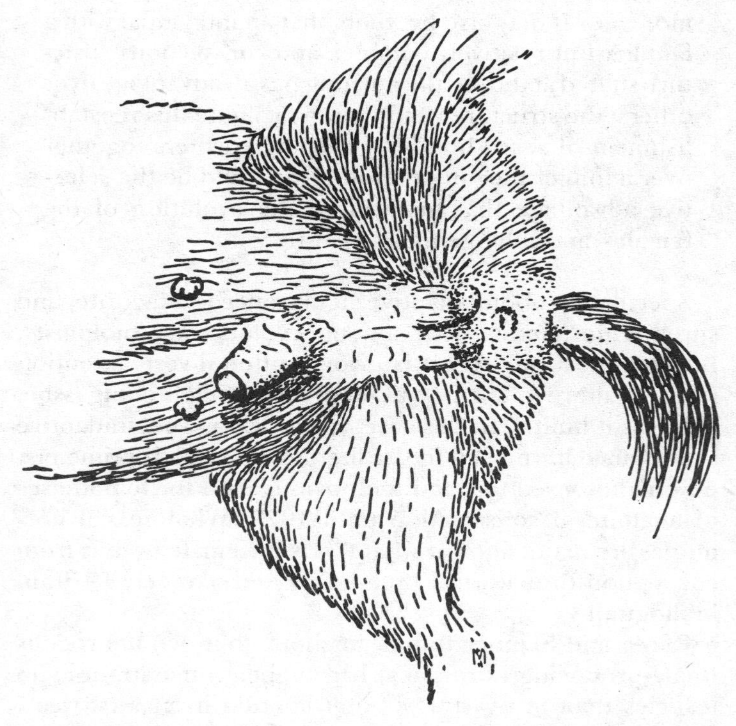

External genitalia of female spotted hyena, showing peniform clitoris and false scrotum. FROM HARRISON MATTHEWS, 1939.

to side, facing in opposite directions. Each lifts its inside hind leg, subordinate individual first, exposing either an erect penis or clitoris, one of the most vulnerable parts of the body, to its partner's teeth. They then sniff and lick each other's genitals for ten to fifteen seconds, primarily at the base of the penis or clitoris and in front of the scrotum or false scrotum.

Kruuk believes that the female clitoris and false scrotum evolved to provide a conspicuous structure serving for recognition in the meeting ceremony. He writes:

> It is impossible to think of any other purpose for this special female feature than for use in the meeting ceremony. . . . It may also be, then, that an individual with a familiar but relatively complex and conspicuous structure sniffed at during the meeting has an advantage over others; the structure would often facilitate this reestablishment of social bonds by keeping partners together over a longer meeting period. This could be the selective advantage that has caused the evolution of the females' and cubs' genital structure.

Speculation about adaptive significance is a favorite, and surely entertaining, ploy among evolutionary biologists. But the question, "What is it for?" often diverts attention from the more mundane but often more enlightening issue, "How is it built?" In this case, speculations about adaptive significance have been in the literature for a long time, yet no one bothered to tread the obvious path for hypotheses of anatomical construction until 1979: What sexual hormones are maintained at what levels by female hyenas from conception to maturity? (See Racey and Skinner, 1979, in bibliography).

Racey and Skinner found, in short, that two androgens (male-producing hormones) had higher concentrations in testicles than in ovaries of adult spotted hyenas (scarcely surprising). Yet, when they investigated levels of the same hormones in blood plasma, they detected *no differences* between males and females. One female contained twin female fetuses, and both had about the same level of testosterone as adult females. Racey and Skinner therefore conclude "that high foetal androgen levels are responsible for the appearance of the male sexual facies in adult female spotted hyenas."

Racey and Skinner affirmed their hypothesis by studying brown and striped hyenas, the other two species of the family Hyaenidae. Neither brown nor striped hyenas develop peniform clitorises or false scrotums. In both species, androgen levels in blood plasma are much lower for females than for males. (Aristotle, by the way, defended hyenas against the charge of hermaphroditism by correctly describing the genitalia of these other species—something of a dodge with respect to the spotted hyena, the source of the legend; but "the master of them that know" was right in any case.)

But why should high levels of androgenic hormones lead to the construction of false penises and scrotums? The animals that form them are still, after all, genetically female. How can female genes produce mimics of male structures, even in a milieu of high androgenic hormones? A look at the developmental basis of sexual anatomy resolves this dilemma.

Mammals share a common pattern for the embryology of sexual organs, and we may therefore use humans as an example. The early embryo is sexually indifferent and contains all precursors and structures necessary for the development of either male or female organs. After about the eighth week following conception, the gonads begin to differentiate as either ovaries or testes. The developing testes secrete androgens, which induce the development of male genitalia. If androgens are absent, or present at low levels, female genitalia are formed.

The internal and external genitalia develop in different ways. For internal genitalia, the early embryo contains precursors of both sexes: the Müllerian ducts (which form the Fallopian tubes and ovaries of females) and the Wolffian ducts (which form the vas deferens—the ducts that carry sperm from the testes to the penis—in males). In females, the Wolffian ducts degenerate and the Müllerian ducts differentiate; males develop by the opposite route.

The external genitalia follow a markedly different pattern. Individuals do not begin with two distinct sets of precursors and then lose one while strengthening the other. Rather, the different organs of male and female develop along diverging routes from the *same* precursor. The male's penis is the same organ as the female's clitoris—they form from the same tissues, are indistinguishable in the early embryo, and follow different pathways later. The male's scrotum is the same organ as the female's labia majora. The two lips simply grow longer, fold over and fuse along the midline, forming the scrotal sac.

The female course of development is, in a sense, biologically intrinsic to all mammals. It is the pattern that unfolds in the absence of any hormonal influence. The male route is a modification induced by secretion of androgens from the developing testes.

The mystery of male mimicry in female hyenas may be solved, in large part, by recognizing these fundamental facts of developmental anatomy. We know from the work of Racey and Skinner that female hyenas maintain high levels of androgenic hormones. We may therefore conclude that the striking and complex peculiarities of sexual anatomy in female spotted hyenas are simply, indeed almost automatically, produced by a single, underlying effect: the secretion of unusually large amounts of androgens by females.

The automatic nature of peniform clitorises and false scrotums in female mammals with high androgen levels can be illustrated by unusual patterns of human development. The adrenal glands also secrete androgens, usually in small amounts. In some genetic females, adrenals are abnormally enlarged and produce high levels of androgens. These baby girls are born with a penis and false scrotum. Several years ago a drug was placed on the market to prevent miscarriages. It had the unfortunate side effect of mimicking the action of natural androgens. Female babies were born with a greatly enlarged clitoris and an empty scrotal sac formed from the fused labia.

I believe that these facts of developmental anatomy must force a revision in the usual interpretation of male mimicry in female spotted hyenas. Evolutionary biologists have too often slipped into a seductively appealing mode of argument about the phenomenon of adaptation. We tend to view every structure as designed for a definite purpose, thus building (in our imagination) a world of perfect design not much different from that concocted by eighteenth-century natural theologians who "proved" God's existence by the perfect architecture of organisms. Adaptationists might allow a little flexibility for tiny and apparently inconsequential structures, but surely anything big, complex, and obviously useful must be built directly by natural selection. Indeed, previous literature on spotted hyenas has assumed that female sexual organs evolved directly for a definite function—as in Kruuk's speculation about the adaptive advantages of conspicuous external genitalia for recognition in the meeting ceremony.

But another scenario is possible and strikes me as more likely. I don't doubt that the basic peculiarity of hyena social organization—the larger size and dominance of females—is an adaptation to something. The easiest pathway to such an adaptation would be a marked rise in the production of androgenic hormones by females (these exist in small amounts in all female mammals). High levels of androgens would entail complex secondary effects as automatic consequences—among them, a peniform clitoris and a false scrotum (we cannot, after all, label the same condition in some abnormal human baby girls as an adaptation). Once these effects are present, some use might be evolved for them—as in the meeting ceremony. But their current utility does not imply that they were built directly by natural selection for the purpose they now serve. (Yes, I know that my scenario might be run in reverse: conspicuous female genitalia are required for the meeting ceremony and are evolved by enhanced androgen levels, thus yielding large female size and dominance as a consequence. I do, however, point out that under our usual preferences for seeing direct adaptation everywhere, my scenario would not even be considered. Indeed, it wasn't in the major works on spotted hyenas.)

We do not inhabit a perfected world where natural selection ruthlessly scrutinizes all organic structures and then molds them for optimal utility. Organisms inherit a body form and a style of embryonic development; these impose constraints upon future change and adaptation. In many cases, evolutionary pathways reflect inherited patterns more than current environmental demands. These inheritances constrain, but they also provide opportunity. A potentially minor genetic change—a rise of androgen level in this case—entails a host of complex, nonadaptive consequences. The primary flexibility of evolution may arise from nonadaptive by-products that occasionally permit organisms to strike out in new and unpredictable directions. What "play" would evolution have if each structure were built for a restricted purpose and could be used for nothing else? How could humans learn to write if our brain had evolved for hunting, social cohesion, or whatever, and could not transcend the adaptive boundaries of its original purpose?

In the second show of his *Cosmos* series, Carl Sagan told the tale of a Japanese crab that carries a portrait of a samurai warrior on its back. He argued that humans have built this face after their own image because local fisherman have been throwing back the most facelike crabs for centuries, thus imposing strong selection pressure for samurai lookalikes (the others get eaten). He used this example as a lead-in for a rapturous discourse on the pervasive power of natural selection.

I doubt this story very much and suspect that the conventional explanation is correct—that the resemblance is accidental and, at best, only slightly strengthened by human intervention. But even if Sagan were right, I believe that he is marveling at the wrong item (or at least failing to give equal time to another remarkable aspect of the case). I am most impressed by a crab's ability to do such an uncrablike thing in the first place—just as the capacity of an inherited developmental system to produce (and so easily) such marked changes in the sexual anatomy of female hyenas grabs me far more than any putative adaptive significance for the change.

The capacity of crabs to make a face on their back did not arise from any selective value such a face might have, since crabs so rarely use this latent ability. Rather, this capacity reflects several deeper facts of crab biology: the bilateral symmetry of the carapace (corresponding by analogy with the bilateral symmetry of the human face), and the fact that many crabs are ornamented by creases along the midline (where a "nose" might form) and perpendicular to it (where "eyes" and "mouths" might be constructed).

The accidental production of a human portrait represents a stunning example of the evolutionary flexibility arising from consequences of an inherited design. Organic material is not putty and natural selection is not omnipotent. Each organic design is pregnant with evolutionary possibilities, but restricted in its paths of potential change. Fishermen might throw back selected starfishes with their five-part symmetry, or snails with their spiral design, for tens of millions of years and never carve a samurai into their hard parts.

Peter Medawar has described science as the "art of the soluble." Evolution might be labeled "the transformation of the possible."

12 Kingdoms without Wheels

SISERA'S MOTHER thought fondly of the booty that her son might bring back—"a prey of divers colors of needlework"—after meeting the armies of Israel led by Deborah and Barak (Judges, chapters 4–5). Yet he was overdue, and she began to worry: "Why tarry the wheels of his chariots?" she inquired anxiously. And rightly did she fear, for Sisera would never return. The Canaanite armies had been routed, while Jael had just transfixed Sisera through the head with a nail (a tent post in modern translations)—ranking her second to Judith among Jewish heroines for the gory dispatch of enemies.

Generals of the biblical armies rode on chariots; their apparatus traveled on carts. But two thousand years later, by the sixth century A.D., the question posed by Sisera's mother could no longer be asked, for wheels virtually disappeared as a means of transportation from Morocco to Afghanistan. They were replaced by camels (Richard W. Bulliet, *The Camel and the Wheel,* 1975).

Bulliet cites several reasons for this counterintuitive switch. The Roman roads had begun to deteriorate and camels were not bound to them. Craftsmanship in harnesses and wagons had suffered a sharp decline. But, most important, camels (as pack animals) were more efficient than carts pulled by draft animals (even by camels). In a long list of reasons for favoring camels to nonmechanized transport by wheels, Bulliet includes their longevity, endurance, ability to ford rivers and traverse rough ground, and savings in manpower (a wagon requires a man for every two animals, but three to six pack camels can be tended by a single person).

We are initially surprised by Bulliet's tale because wheels have come to symbolize in our culture the sine qua non of intelligent exploitation and technological progress. Once invented, their superiority cannot be gainsaid or superseded. Indeed, "reinventing the wheel" has become our standard metaphor for deriding the repetition of such obvious truths. In an earlier era of triumphant social Darwinism, wheels stood as an ineluctable stage of human progress. The "inferior" cultures of Africa slid to defeat; their conquerors rolled to victory. The "advanced" cultures of Mexico and Peru might have repulsed Cortés and Pizarro if only a clever artisan had thought of turning a calendar stone into a cartwheel. The notion that carts could ever be replaced by pack animals strikes us not only as backward but almost sacrilegious.

The success of camels reemphasizes a fundamental theme of these essays. Adaptation, be it biological or cultural, represents a better fit to specific, local environments, not an inevitable stage in a ladder of progress. Wheels were a formidable invention, and their uses are manifold (potters and millers did not abandon them, even when cartwrights were eclipsed). But camels may work better in some circumstances. Wheels, like wings, fins, and brains, are exquisite devices for certain purposes, not signs of intrinsic superiority.

The haughty camel may provide enough embarrassment for any modern Ezekiel, yet this column might seem to represent still another blot on the wheel's reputation (though it does not). For I wish to pose another question that seems to limit the wheel. So much of human technology arose by recreating the good designs of organisms. If art mirrors nature and if wheels are so successful an invention, why do animals walk, fly, swim, leap, slither, and creep, but never roll (at least not on wheels)? It is bad enough that wheels, as human artifacts, are not always superior to nature's handiwork. Why has nature, so multifarious in her ways, shunned the wheel as well? Are wheels a poor or rarely efficient way to make progress after all?

In this case, however, the limit lies with animals, not with the efficiency of wheels. A vulgarization of evolution, presented in many popular accounts, casts natural selection as a perfecting principle, so accurate in its operation, so unconstrained in its action, that animals come to embody a set of engineering blueprints for optimal form (see essay 11). Instead of replacing the older "argument from design"—the notion that God's existence can be proved by the harmonies of nature and the clever construction of organisms—natural selection slips into God's old role as perfecting principle.

But the proof that evolution, and not the fiat of a rational agent, has built organisms lies in the imperfections that record a *history* of descent and refute creation from nothing. Animals cannot evolve many advantageous forms because inherited architectural patterns preclude them. Wheels are not flawed as modes of transport; I am sure that many animals would do far better with them. (The one creature clever enough to build them, after all, has gotten some mileage from the invention, the superiority of camels in certain circumstances notwithstanding.) But animals cannot construct wheels from the parts that nature provides.

As its basic structural principle, a true wheel must spin freely without physical fusion to the solid object it drives. If wheel and object are physically linked, then the wheel cannot turn freely for very long and must rotate back, lest connecting elements be ruptured by the accumulated stress. But animals must maintain physical connections between their parts. If the ends of our legs were axles and our feet were wheels, how could blood, nutrients, and nerve impulses cross the gap to nurture and direct the moving parts of our natural roller skates? The bones of our arms may be unconnected, but we need the surrounding envelopes of muscle, blood vessels, and skin—and therefore cannot rotate our arms even once around our shoulders.

We study animals to illuminate or exemplify nature's laws. The highest principle of all may be nature's equivalent of the axiom that for every hard-won and comforting regularity, we can find an exception. Sure enough—somebody out there has a wheel. In fact, at this very moment, wheels are rotating by the millions in your own gut.

Escherichia coli, the common bacillus of the human gut, is about two micrometers long (a micrometer is one-thousandth of a millimeter). Propelled by long whiplike threads called flagella (singular, flagellum), an *E. coli* can swim about ten times its own length in a second. Lest swimming seem easy for a creature virtually unaffected by gravitational forces and moving through a supporting and easily yielding fluid, I caution against extrapolating our view to a bacterium's world. The perceived viscosity of a fluid depends upon an organism's dimensions. Decrease a creature's size and water quickly turns to molasses. Howard C. Berg, the Colorado biologist who demonstrated how flagella operate, compares a bacterium moving in water to a man trying to swim through asphalt. A bacterium cannot coast. If its flagella stop moving, a bacterium comes to an abrupt halt within about a millionth of its body length. The flagella work wonderfully well in trying circumstances.

After Berg had modified his microscope to track individual bacteria, he noted that an *E. coli* moves in two ways. It may "run," swimming steadily for a time in a straight or slightly curved path. Then it stops abruptly and jiggles about—a "twiddle" in Berg's terminology. After twiddling, it runs off again in another direction. Twiddles last a tenth of a second and occur on an average of once a second. The timing of twiddles and the directions of new runs seem to be random unless a chemical attractant exists at high concentration in one part of the medium. A bacterium will then move up-gradient toward the attractant by decreasing the probability of twiddling when a random run carries it in the right direction. When a random run moves in the wrong direction, twiddling frequency remains at its normal, higher level. The bacteria therefore drift toward an attractant by increasing the lengths of runs in favored directions.

The bacterial flagellum is built in three parts: a long helical filament, a short segment (called a hook) connecting the filament to the flagellar base, and a basal structure embedded in the cell wall. Biologists have argued about how bacteria move since Leeuwenhoek first saw them in 1676. Most models assumed that flagella are fixed rigidly to the cell wall and that they propel bacteria by waving to and fro. When such models had little success in explaining the rapid transition between runs and twiddles, some biologists suggested that flagella might tag passively along and that some other (and unknown) mechanism might move bacteria.

Berg's observations revealed something surprising, hinted at and proposed in theory before, but never adequately demonstrated: the bacterial flagellum operates as a wheel. It rotates rigidly like a propeller, driven by a rotatory "motor" in the basal portion embedded in the cell wall. Moreover, the motor is reversible. *E. coli* runs by rotating the flagella in one direction; it twiddles by abruptly stopping and rotating the flagella the other way!

Berg could observe the rotation and correlate its direction with runs and twiddles by following free-swimming bacteria in his machine, but S. H. Larsen and others, working in Julius Adler's laboratory at the University of Wisconsin, provided an even more striking demonstration. They isolated two mutant strains of *E. coli*—one that runs and never twiddles and another that twiddles incessantly. They "tethered" these mutant bacteria to glass slides, using antibodies that attach either to the hook or filament of the flagella and also, fortunately, to glass. Thus, the bacteria are affixed to the slide by their flagella. Larsen noted that the tethered bacteria rotate continually about their immobilized flagella. The running mutants turn counterclockwise (as viewed from outside the cell), while the twiddling mutants turn clockwise. The flagellar wheel has a reversible motor.

The biochemical basis of rotation has not yet been elucidated, but the morphology can be resolved. Berg proposes that the bottom end of the flagellum expands out to form a thin ring rotating freely in the cytoplasmic membrane of the cell wall. Just above, another ring surrounds the flagellar base, without attaching to it. This second ring is mounted rigidly on the cell wall. The lower ring (and entire flagellum) rotates freely, held in position by the surrounding upper ring and the cell wall itself.

Some exceptions in nature are dispiriting—the nasty, ugly, little facts that spoil great theories, in Huxley's aphorism. Others are enlightening and serve only to reinforce a regularity by identifying both its scope and its reasons. These are the exceptions that prove (or probe) rules—and the flagellar wheel falls into this happy class.

Is it accidental that wheels only occur in nature's smallest creatures? Organic wheels require that two parts be juxtaposed without physical connection. I argued previously that this cannot be accomplished in creatures familiar to us because connection between parts is an integral property of living systems. Substances and impulses must be able to move from one segment to another. Yet, in the smallest organisms—and in them alone—substances can move between two unconnected parts by diffusing through membranes. Thus, single cells, including all of ours of course, contain organelles lying within the cytoplasm and communicating with other parts of the cell, not by physical connection, but by passage of molecules through bounding membranes. Such structures could, in principle, be designed to rotate like wheels.

The principle that restricts such communication without physical connection to the smallest organisms (or to similarly sized parts of larger organisms) embodies a theme that has circulated extensively throughout these essays (see sections in *Ever Since Darwin* and *The Panda's Thumb*): the correlation of size and shape through the changing relationship of surfaces and volumes. With surfaces ($length^2$) increasing so much more slowly than volumes ($length^3$) as an object grows, any process regulated by surfaces but essential to volumes must become less efficient unless the enlarging object changes its shape to produce more surface. The external boundary is surface enough for communication between the organelles of a single cell with their minuscule volumes. But the surface of a wheel as large as a human foot could not provision the wheelful of organic matter within.

Large organisms must evolve channels—physical connections—to convey the nutrients and oxygen that can no longer diffuse through external surfaces.

Wheels work well, but animals are debarred from building them by structural constraints inherited as an evolutionary legacy. Adaptation does not follow the blueprints of a perfect engineer. It must work with parts available. Yet when I survey animals in all their stunning, if wheel-less, variety, I can only marvel at the diversity and good design that a few basic and highly constrained organic patterns have produced. Forced to make do, we do rather well.

Postscript

I did not know how many artists and writers of fiction had made up for nature's limitations until readers began to submit their favorite stories. To choose just one example in each category, G. W. Chandler told me that one of the *Oz* novels featured some four-legged rollers known as wheelers. They were, in fact, built in just the way I argued an animal could not work—with wheels for feet and the ends of legs for axles. D. Roper sent me a print of M. C. Escher's "curl-up," a lithograph showing hundreds of curious creatures wandering through a typical Escher landscape of impossible staircases. They climb by dragging a segmented body along on three pairs of humanoid legs. When they hit a flat surface, they roll up and roll along. These, of course, are permissible "one part" wheels, (like tumbling tumbleweeds), not the impossible wheel and axle combination. Still, Escher specifically created them to make up for nature's limitation since he writes that the lithograph was inspired by his "dissatisfaction concerning nature's lack of any wheelshaped living creatures. . . . So the little animal shown here . . . is an attempt to fill a long-felt want."

Still, as usual, nature wins again. Robert LaPorta and Joseph Frankel both wrote to tell me that I had missed another of nature's real wheels. They directed me to the work of Sidney Tamm, which, I am ashamed to say, I did not know when I wrote the original article. Dr. Tamm has found wheels in single-celled creatures that live in the guts of termites. They therefore (whew!) fall into the category of permissible exceptions at small dimensions.

The body of this protist contains an axostyle (a kind of axis running the length of its body) that rotates continuously in one direction. The organelles of the anterior end (including the nucleus) are attached to the axostyle and rotate with it—"much like turning a lollipop by the stick," as Tamm notes. But, and we now encounter the more curious and wheel-like point, the entire anterior end, including the cell surface, rotates along with the axostyle relative to the rest of the body.

Tamm demonstrated this peculiar motion with an ingenious experiment in which he attached small bacteria all over the cell's outer surface. Those attached to the front end rotated continuously with respect to those adhering to the back end. But bacteria did not attach to a narrow band between front and back, and this band must therefore represent a zone of shear. Tamm then studied the structure of the cell-membrane by freeze-fracture electron microscopy and found it to be continuous across the shear zone. Tamm concludes that the entire surface must be fluid and that shear zones could, in theory, form anywhere upon it. A very strange creature! "Prais'd be the fathomless universe," Whitman wrote, "for life and joy, and for objects and knowledge curious."

13 What Happens to Bodies if Genes Act for Themselves?

THE UNCOMMON good prose of scientists is more often spare than flowery. In my favorite example, James D. Watson and Francis Crick used less than a page to announce their structure of DNA in 1953. They began with the sparsest announcement: "We wish to suggest a structure for the salt of deoxyribose nucleic acid (D.N.A.). This structure has novel features which are of considerable biological interest." And they ended with a reminder that they had not overlooked a major point just because they had chosen to defer its discussion: "It has not escaped our notice that the specific pairing we have postulated immediately suggests a possible copying mechanism for the genetic material" (that is, the two strands of the double helix would pull apart and each then act as a template for the reconstitution of its partner).

Francis Crick, now a professor at the Salk Institute in southern California, has continued to generate controversial, challenging hypotheses (and he has often been right). In late 1981, he published a book, *Life Itself,* advocating a theory of "directed panspermia"—the idea that Earth's original life arrived as microorganisms dispatched by intelligent beings who chose not to make the long journey themselves. (Ten will get you fifty that he's wrong this time—but only fifty; he's been right too often.)

Crick has also not lost his gift for a well-turned phrase. In the presentation of his latest controversial hypothesis, published in *Nature* (April 17, 1980) with Salk colleague Leslie Orgel as first author, he outdid the last line of his 1953 paper with Watson. Orgel and Crick conclude: "The main facts are, at first sight, so odd that only a somewhat unconventional idea is likely to explain them." Indeed, the facts are so interesting, and the wondering about them so intense, that the same issue of *Nature* carried an accompanying article by Dalhousie University biologists W. Ford Doolittle and Carmen Sapienza, who had, quite independently, devised the same explanation and argued the case, in many ways, more forcefully.

What, then, are these disturbing facts? When a younger Crick determined the structure of DNA in 1953, and others cracked the genetic code a few years later, everything seemed momentarily to fall into order. The old idea of genes as beads on a string (the chromosome) seemed to gain its vindication from the Watson-Crick model. Each of the three nucleotides in DNA codes for an amino acid (via an RNA intermediary); a string of amino acids makes a protein. Perhaps we could simply read down a chromosome to find genes lined up, one after the other, each ready to begin the assembly of its essential part.

It was not to be so. Is it ever? We now know that the genetic material of higher organisms is vastly more complex. Many genes come in pieces, separated in DNA by sequences of nucleotides that are not transcribed into RNA. Many proteins are coded by partial sequences on two or more chromosomes. What controls regulate their assembly? (Human globin, the protein component of hemoglobin, contains alpha and beta chains—and the genes for each chain are on separate chromosomes.)

Even more disturbing (and exhilarating) is the discovery, made more than a decade ago but gathering intensity ever since, that only a small percentage of DNA codes for proteins in higher organisms—and that these are the only bits of DNA whose function we may truly understand at the moment. In humans, somewhat more than 1 percent, but not as much as 2 percent, of DNA codes for proteins. Much of the rest contains sequences that are repeated over and over again—hundreds or thousands of identical (or nearly identical) beads, sometimes following one after the other, but sometimes dispersed widely over several chromosomes. Why so many copies? What do they do? The "selfish DNA" hypothesis of Doolittle, Sapienza, Orgel, and Crick provides an unusual answer to the puzzling question of why so much DNA exists in repeated copies (but I will keep you in suspense for a bit and discuss the conventional answers first).

Higher organisms contain different classes of repeated DNA. One type, called highly repeated or satellite DNA, contains short and simple sequences repeated hundreds of thousands or millions of times; 5 percent or so of human DNA falls into this class. We hardly have a clue about the origin and function of satellite DNA; neither the selfish DNA hypothesis nor the conventional hypotheses can explain it. Satellite DNA is, as they say, a "whole 'nother" story waiting to be told.

The current debate over the conventional and selfish DNA hypotheses centers upon the so-called intermediate or middle-repetitive DNA, some 15 to 30 percent of both the human and the fruit fly genome. Middle-repetitive DNA exists in tens to a few hundred copies per sequence; the copies are often widely dispersed on several chromosomes.

I have said nothing, so far, about the DNA of simpler organisms—the prokaryotic bacteria and blue-green algae, which have no nucleus and carry their DNA in a single chromosome. The DNA of prokaryote (prenucleate) organisms is "better behaved" with reference to the original hopes of the Watson-Crick model. Most bacterial DNA is single copy and protein coding, almost those beads on a string after all. But even prokaryotes are not immune to repetition. A hot topic of late concerns the presence in prokaryotes of so-called transposons, transposable elements, or more colorfully, jumping genes. These sequences of DNA, as their various names proclaim, can repeat themselves and then autonomously move about to other positions on the bacterial chromosome. They often exist in about as many copies as middle-repetitive DNA in eukaryotes (higher organisms with a nucleus and paired chromosomes). This has led many biologists to propose that at least some of the middle-repetitive DNA in higher organisms amplifies itself by the same mechanism of transposition. (The selfish DNA hypothesis assumes a correspondence between prokaryote transposons and the source of middle-repetitive DNA in eukaryotes. Some middle-repetitive DNA probably arises in other ways, and selfish DNA will therefore not explain all of it.)

Conventional arguments for the existence of middle-repetitive DNA follow the usual Darwinian perspective. Evolution is about the struggle of organisms to leave more surviving offspring in future generations. This struggle operates by natural selection and selection is a potent editor. Major features of organisms—and some 25 percent of the genetic material cannot be minor—must exist because they provide some advantages to organisms in the struggle for life. We must, in other words, find a function for middle-repetitive DNA in terms of advantages to the bodies that carry it.

Rumblings of claims for nonadaptive and nonfunctional status have been heard from time to time (selfish DNA is the first, and more subtle, explosion for this perspective). Still, as Doolittle and Sapienza detail in their article, the overwhelming majority of proposals have hewed to Darwinian orthodoxy: they assume that middle-repetitive DNA cannot exist in such amounts unless it confers direct adaptive benefits upon organisms. (I will save myself some words from now on by simply writing "repetitive DNA" when I mean "middle-repetitive DNA" only.)

The conventional adaptationist hypotheses have fallen into two classes: one, I believe, obviously wrong on (unrecognized) principle; the other undoubtedly correct in part (I do not believe that all repetitive DNA is selfish DNA). The unreasonable arguments postulate what I like to call a "retrospective significance" for repetitive DNA—that is, they justify its existence by discussing the benefits it may confer upon distant evolutionary futures.

Suppose all working genes could only exist in one copy

that coded for an essential protein. How then could substantial evolutionary change ever occur? What will supply the essential protein while evolution monkeys about with the only coding sequence that produces it? But if a gene can repeat itself, then one copy might continue to code for the essential protein, leaving the other free to change. Thus, potential flexibility for evolutionary change has often been cited as the primary significance of repetitive DNA.

I have no quarrel with the idea that redundancy may supply the flexibility that evolution requires for initiating major changes. Susumu Ohno, who first popularized this idea in 1970 in a brilliant book *(Evolution by Gene Duplication),* argued that, without redundancy "from a bacterium only numerous forms of bacteria would have emerged." Duplication supplies the raw material of major evolutionary change: "The creation of a new gene from a redundant copy of an old gene is the most important role that gene duplication played in evolution."

But think about it for a moment. The argument is sound and may represent, in fact, the major *effect* of gene duplication for evolution. Yet unless our usual ideas about causality are running in the wrong direction, this flexibility simply cannot be the adaptive explanation for why repetitive DNA exists. Selection works for the moment. It cannot sense what may be of use ten million years hence in a distant descendant. The duplicated gene may make future evolutionary change possible, but selection cannot preserve it unless it confers an "immediate significance." Future utility is an important consideration in evolution, but it cannot be the explanation for current preservation. Future utilities can only be the *fortuitous effects* of other direct reasons for immediate favor.

(The confusion of *current utility* with *reasons for past historical origin* is a logical trap that has plagued evolutionary thinking from the start—see essay 11. Feathers work beautifully in flight, but the ancestors of birds must have evolved them for another reason—probably for thermoregulation—since a few feathers on the arm of a small running reptile will not induce takeoff. Our brains enlarged for a set of complex reasons, but surely not so that some of us could write essays about it. Interested readers may wish to consult a technical article that Elisabeth Vrba and I have written about this subject—see bibliography. We wish to restrict the term *adaptation* only to those structures that evolved for their current utility; those useful structures that arose for other reasons or for no conventional reason at all, and were then fortuitously available for other usages, we call exaptations. New and important genes that evolved from a repeated copy of an ancestral gene are partial exaptations, for their new usage cannot be the reason for the original duplication.)

The second set of adaptive arguments is legitimate in proposing an immediate selective benefit for repeated DNA. If genes move about and insert themselves on different chromosomes, for example, they may occasionally link up with other segments of DNA to form advantageous new combinations. More importantly, much DNA, while not coding for protein itself, may play a role in regulating the DNA that does. This regulatory DNA may turn other genes on and off and may determine the sequence and location of expression for the genes that do code for proteins. If repetitive DNA performs these regulatory functions, then its dispersal throughout the genome can have profound immediate effects. Inserted into a new chromosome, it may turn adjacent genes on and off in new ways and sequences. It may, for example, bring together the products of two genes that had never been in proximity. This new combination may benefit an organism (see the classic article of Roy Britten and Eric Davidson, 1971).

Yet, for all these efforts, the nagging suspicion remains that these adaptive explanations cannot account for all repetitive DNA. There is simply too much of it, too randomly dispersed, too apparently nonsensical in its construction, to argue that each item perseveres because natural selection has favored it in a regulatory role. The selfish DNA hypothesis proposes a fundamentally different explanation for much of this repetition. It is radical in that literal sense of getting to the roots, for it demands that we reassess some basic and usually unquestioned assumptions of evolutionary argument—what Orgel and Crick meant when they spoke of facts "so odd that only a somewhat unconventional idea is likely to explain them."

The argument is simplicity itself once you establish the frame of mind to permit it: if repetitive DNA is transposable, then why do we need an adaptive explanation for it at all (at least in conventional terms of benefits to bodies)? It may simply spread of its own accord from chromosome to chromosome, making more copies of itself while other "sedentary" genes cannot. These extra copies may persist, not because they confer advantages upon bodies, but for precisely the opposite reason—because bodies do not notice them. If they have no effect upon bodies, if they are (in this sense) "junk," then what is to stop their spread? They are merely playing Darwin's game, but at the "wrong" level. We usually think of natural selection as a struggle among bodies to leave more surviving offspring. Here certain genes have found a way, through transposability, or "jumping," to leave more copies of themselves *within* a body. Is any other explanation required? Orgel and Crick's title reflects this reversed perspective: "Selfish DNA: The Ultimate Parasite."

I can now almost hear the disappointment and anger of some readers: "That bastard Gould. He led us along for pages, and now he gives an explanation that is no explanation at all. It just plain happens, and that's all there is to it. Is this a joke or a counsel of despair?" I beg to differ from this not entirely hypothetical adversary (a composite constructed from several real responses I have received to verbal descriptions of the selfish DNA hypothesis). The explanation seems hokey only in the context of adherence to traditional views that all important features must be adaptations and that bodies are *the* agent of Darwinian processes. The radical content of selfish DNA is not the explanation itself, but the reformulated perspective that must be assimilated before the explanation confers any satisfaction.

If bodies are the only "individuals" that count in evolution, then selfish DNA is unsatisfying because it does nothing for bodies and can only be seen as random with respect to bodies. But why should bodies occupy such a central and privileged position in evolutionary theory? To be sure, selection can only work on discrete individuals with inherited continuity from ancestor to descendant. But are bodies the only kind of legitimate individuals in biology? Might there not be a hierarchy of individuals, with legitimate categories both above and below bodies: genes below, species above? (I confess to what evolutionists call a "preadaptation" for favorable response to the selfish DNA hypothesis. I have long argued that species must be viewed as true evolutionary units and that macroevolutionary trends are often powered by a "species selection" that is analogous to, but not identical with, natural selection acting upon bodies.) Selfish DNA may do nothing for bodies, but bodies are the wrong level of analysis. From a gene's point of view, transposable elements have developed a great Darwinian innovation: they have found a way to make more surviving copies of themselves (by repetition and transposition), and this, in itself, is the evolutionary *summum bonum.* If bodies don't notice this repetition, and therefore cannot suppress it by dying or failing to reproduce, then so much the better for repeating genes.

In this sense, selfish DNA is about the worst possible name for the phenomenon, for it records the very prejudice that the new structure of explanation should be combating: an exclusive focus on bodies as evolutionary agents. When we call repetitive DNA "selfish," we imply that it is acting for itself when it should be doing something else, namely, helping bodies in their evolutionary struggle. Likewise, we should not refer to repetitive DNA as "nonadaptive," for although it may not be helping bodies, it is acting as its own Darwinian agent. I can't think of a much better name in a language replete with anthropocentric terms, but how about "self-centered DNA"—without the opprobrious overtones that "selfish" inevitably contains.

Another argument against the use of selfish DNA lies in its historical source: Richard Dawkins's book *The Selfish Gene* (1976). Dawkins argued that bodies are the wrong level of evolutionary analysis and that all evolution is nothing but a struggle among genes. Bodies are merely temporary containers for their selfish genes. Superficially, this looks like selfish DNA writ larger, hence Orgel and Crick's decision to borrow the term. In fact, the theories of selfish genes and selfish DNA could not be more different in the structures of explanation that nurture them.

Dawkins writes as a strict Darwinian, committed to the idea that all features must be interpreted as adaptations and that all of evolution is a struggle for existence among individuals at the lowest level. He merely decided that Darwinians weren't radical enough in reducing such higher-level reveries as "the good of the species" or "the harmony of nature" to the unrestrained struggle of organisms. The struggling items are one level lower—genes rather than bodies—and the Darwinian program of reduction can go even further than modern supporters had dared to hope.

Selfish DNA, on the other hand, gains its rationale from the antireductionistic belief that evolution works on a hierarchy of legitimate levels that cannot be collapsed to the first rung of the scale. Dawkins's selfish genes increase in frequency because they have effects upon bodies, aiding them in their struggle for existence. Selfish DNA increases in frequency for precisely the opposite reason—because it initially has *no effect* on bodies and therefore is not suppressed at this legitimate higher level. Dawkins's theory is an unconventional proposal to explain ordinary adaptation of bodies (see my critique in *The Panda's Thumb*). Selfish DNA survives only because it makes no difference to bodies.

But if middle-repetitive DNA is self-centered, why does it only exist in hundreds of copies within genomes? If it can spread by transposition while other genes cannot, why does it not generate millions and billions of copies, eventually crowding everything else out? What stops it? Why is it behaving as an "intelligent" parasite (enough copies to be comfortable and powerful, but not enough to destroy the host and itself), rather than as a voracious cancer?

The potential answer to this question, proposed by both sets of authors, illustrates another interesting point about the hierarchical mode of thinking that underlies the theory of self-centered DNA. In hierarchical models, levels are not independent, walled off by impenetrable boundaries from those above and below. Levels leak and interact. Arthur Koestler, whom I do not usually praise but whose commitment to hierarchy I find admirable, chose as his metaphor for hierarchy the double-faced god Janus, standing at one level but looking for connections in both directions.

Consider different forms of selection working at levels of gene, body, and species. A transposon enters a genetic system and begins to amplify itself by replication and movement. In the process of selection among genes, it is increasing by an analog of what we would call "differential birth" in natural selection among bodies. Its increase initially produces no interaction with the level of natural selection upon bodies, and nothing suppresses its intrinsic drive to manufacture more copies.

But eventually, if its increase continues unabated, bodies must begin to notice. There is an energetic cost attached to the replication, generation after generation, of hundreds or thousands of copies of DNA sequences that do nothing for the bodies investing that energy. Bodies may not notice a few copies, but vast numbers must eventually produce a disadvantage at the good old Darwinian level of natural selection among bodies. At this point, a further increase in self-centered DNA will be suppressed because bodies carrying too many copies will suffer in natural selection, taking all their copies with them when they die or fail to reproduce. The usual level of tens to hundreds of copies may well represent a balance between inexorable increase at the level of selection among genes and eventual suppression at the next level of selection among bodies. Levels are connected by complex ties of feedback. My plea for a recognition of selection at levels other than bodies is not a negation of Darwinian theory but an attempt to enrich it.

The arguments will continue for a long time. One group of scientists notes the similarity in arrangement within chromosomes of repetitive sequences in two creatures as evolutionarily distant as the toad *Xenopus laevis* and the sea urchin *Strongylocentrotus purpuratus.* This similarity refutes self-centered DNA and points to common function, since wandering transposons, beholden only to their own level, should disperse more randomly among chromosomes. Others point out that an important transposable element in yeast and another in the fruit fly *Drosophila melanogaster* are represented in different strains of the same species by about the same number of copies, but in very different positions among chromosomes. Do the different positions represent self-centered amplification and the similarity in numbers reflect suppression at the higher level of selection upon bodies?

As with all interesting questions in natural history, the solution requires an inquiry about relative frequency, not an absolute yes or no. The logic of self-centered DNA seems sound. The question remains: how important is it? How much repetitive DNA is self-centered DNA? If the answer is "way less than one percent" because conventional selection on bodies almost always overwhelms selection among genes, then self-centered DNA is one more good and plausible idea scorned by nature. If the answer is "lots of it," then we will need a fully articulated hierarchical theory of evolution. My own inclinations are, obviously, for hierarchy. Cartesian reductionism has been the source of science's triumph for 300 years; but I suspect that we have reached its limits in several areas.

We have legitimate, idiosyncratic reasons for continuing our linguistic habit of identifying "individuals" with bodies, and for granting a primacy to bodies among the objects of nature. I can't, for example, imagine any acceptable politics that does not focus upon the primacy of individual bodies—and we weep for the inhumanity of those that did not, but flourished for a time nonetheless. Nature, however, acknowledges many kinds of individuals, both great and small.

14 Hen's Teeth and Horse's Toes

VANITY LICENSE PLATES are the latest expression of an old conviction that distinctive conveyances reflect status or, at least, compel notice. We can build our modern machines to order, but nature has narrower limits. Horses of unusual size or color commanded great favor, but Julius Caesar ventured beyond the mere accentuation of normality in choosing his favorite mount. The historian Suetonius writes that Caesar

> used to ride a remarkable horse, which had feet that were almost human, the hoofs being cleft like toes. It was born in his own stables, and as the soothsayers declared that it showed its owner would be lord of the world, he reared it with great care, and was the first to mount it; it would allow no other rider.

Normal horses represent the limit of evolutionary trends for the reduction of toes. Ancestral *Hyracotherium* (popularly, but incorrectly, known as *Eohippus*) had four toes in front and three in back, while some earlier forebear undoubtedly possessed the original mammalian complement of five on each foot. Modern horses retain but a single toe, the third of an original five. They also develop vestiges of the old second and fourth toes as short splints of bone, mounted high and inconspicuously above the hoof.

Abnormal horses with extra digits have been admired and

studied since Caesar's time. O. C. Marsh, a founder of vertebrate paleontology in America, took a special interest in these aberrant animals and published a long article on "Recent polydactyle horses" in April 1892. Marsh had two major claims upon fame, one dubious—his acrimonious battles with E. D. Cope in collecting and describing vertebrate fossils from the American West—and one unambiguous—his success in deciphering the evolution of horses, the first

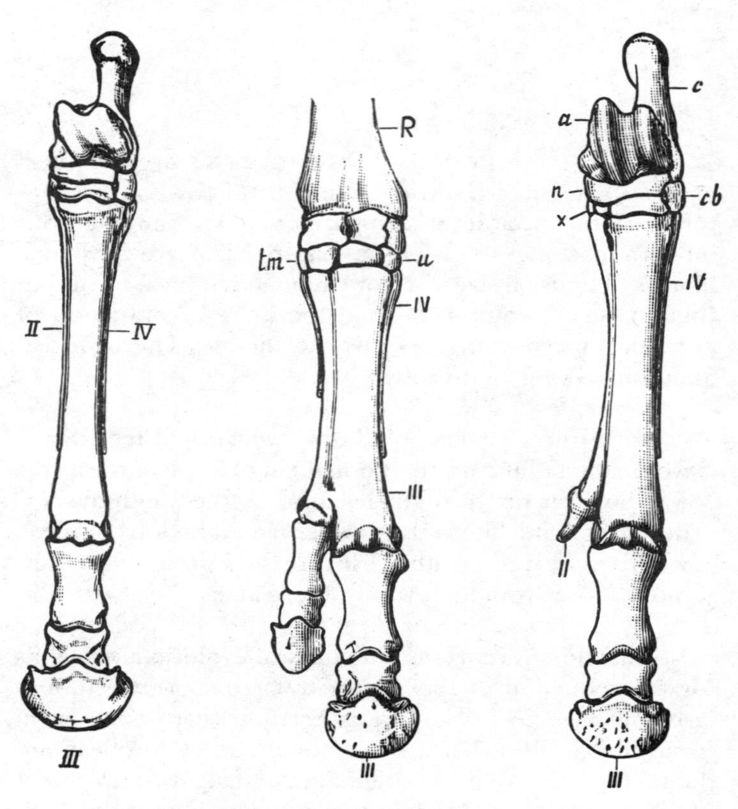

Marsh's 1892 figures of polydactyl horses. Left, a normal horse. Note the splint remnants of side toes labeled II and IV. Middle: polydactyly by duplication. The side splints are still present and the extra toe is a duplicated third digit. Right: polydactyly by atavism. The extra toe is an enlarged side splint.

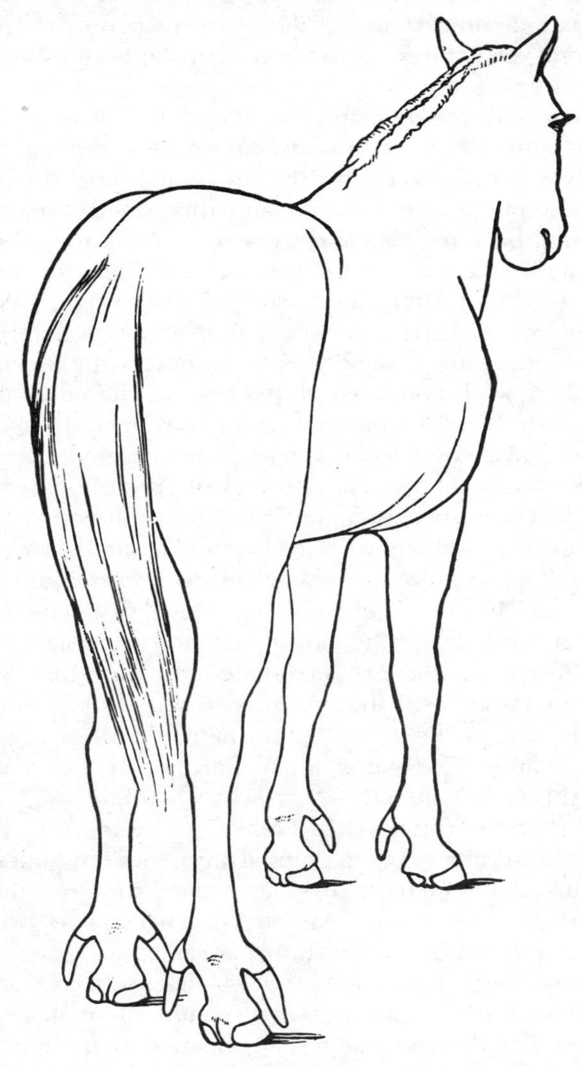

Marsh's 1892 figure of a polydactyl horse, the "horned horse from Texas."

adequate demonstration of descent provided by the fossil record of vertebrates, and an important support in Darwin's early battles.

Marsh was puzzled and fascinated by these aberrant horses with extra toes. In most cases, the additional toe is merely a duplicate copy of the functional third digit. But Marsh found that many two- and three-toed horses had harkened back to their ancestors by developing either or both of the side splints into functional (or nearly functional) hoofed toes. (A later, and particularly thorough, German monograph of 1918 concluded that about two-thirds of horses with extra toes had simply duplicated the functional third digit, while about one-third had resuscitated an ancestral feature by developing the vestigial splints of their second or fourth toe into complete, hoofed digits.)

These apparent reversions to previous evolutionary states are called atavisms, after the Latin *atavus:* literally, great-great-great-grandfather; more generally, simply ancestor. The biological literature is studded with examples of the genre, but they have generally been treated anecdotally as mere curiosities bearing no important evolutionary message. If anything, they are surrounded with the odor of slight embarrassment, as if the progressive process of evolution did not care to be reminded so palpably of its previous imperfections. The synonyms of European colleagues express this feeling directly—"throwback" in England, *pas-en-arrière* ("backward step") in France, and *Rückschlag* ("set-back") in Germany. When granted any general significance, atavisms have been treated as marks of constraint, as indications that an organism's past lurks just below its present surface and can hold back its future advance.

I would suggest an opposite view—that atavisms teach an important lesson about potential results of small genetic changes, and that they suggest an unconventional approach to the problem of major transitions in evolution. In the traditional view, major transitions are a summation of the small changes that adapt populations ever more finely to their local environments. Several evolutionists, myself included, have become dissatisfied with this vision of smooth extrapolation. Must one group always evolve from another through an insensibly graded series of intermediate forms? Must evolution proceed gene by gene, each tiny change producing a correspondingly small alteration of external appearance? The fossil record rarely records smooth transitions, and it is often difficult even to imagine a function for all hypothetical intermediates between ancestors and their highly modified descendants.

One promising solution to this dilemma recognizes that certain kinds of small genetic changes may have major, discontinuous effects upon morphology. We can make no one-to-one translation between extent of genetic change and degree of alteration in external form. Genes are not attached to independent bits of the body, each responsible for building one small item. Genetic systems are arranged hierarchically; controllers and master switches often activate large blocks of genes. Small changes in the timing of action for these controllers often translate into major and discontinuous alterations of external form. Most dramatic are the so-called homeotic mutants discussed in the following essay.

The current challenge to traditional gradualistic accounts of evolutionary transitions will take root only if genetic systems contain extensive, hidden capacities for expressing small changes as large effects. Atavisms provide the most striking demonstration of this principle that I know. If genetic systems were beanbags of independent items, each responsible for building a single part of the body, then evolutionary change could only occur piece by piece. But genetic systems are integrated products of an organism's history, and they retain extensive, latent capacities that can often be released by small changes. Horses have never lost the genetic information for producing side toes even though their ancestors settled on a single toe several million years ago. What else might their genetic system maintain, normally unexpressed, but able to serve, if activated, as a possible focus for major and rapid evolutionary change? Atavisms reflect the enormous, latent capacity of genetic systems, not primarily the constraints and limitations imposed by an organism's past.

My latent interest in atavism was recently kindled by a report of something that has no right to exist if one of our most venerable similes expresses literal truth—hen's teeth. On February 29, 1980 (enough of a rarity in itself), E. J. Kollar and C. Fisher reported an ingenious technique for coaxing chickens to reveal some surprising genetic flexibility retained from a distant past.

They took epithelial (outer) tissue from the first and second gill arches of a five-day-old chick embryo and combined it with mesenchyme (inner embryonic tissue) of sixteen- to eighteen-day-old mouse embryos taken from the region where first molar teeth form. A fascinating evolutionary tale lies hidden in this simple statement as well. Jaws evolved from bones supporting the anterior gills of ancestral fishes. All vertebrate embryos still develop the anterior gill arches first (as ancestral embryos did) and then transform them during development into jaws (as ancestors did not in retaining the forward gills throughout life). Thus, if the embryonic tissues of chickens still retain any capacity for forming teeth, the epithelium of the anterior gill arches is the place to look.

Kollar and Fisher took the combined embryonic tissue of mouse and chicken and grew it in what might strike readers as a bizarre and unlikely place—the anterior chambers of the eyes of adult nude mice (but where else in an animal's body can one find an open space, filled with liquid that is not circulating?). In ordinary teeth, made by a single animal, the outer enamel layer forms from epithelial tissue and the underlying dentin and bone from mesenchyme. But mesenchyme cannot form dentin (although it can produce bone) unless it can interact directly with epithelium destined to form enamel. (In embryological jargon, epithelium is a necessary inducer, although only mesenchyme can form dentin.)

When Kollar and Fisher grafted mouse mesenchyme alone into the eyes of their experimental animals, no dentin developed, but only spongy bone—the normal product of mesenchyme when deprived of contact with enamel epithelium as an inducer. But among fifty-five combined grafts of mouse mesenchyme and chick epithelium, ten produced dentin. Thus, chick epithelium is still capable of inducing mesenchyme (from another species in another vertebrate class yet!) to form dentin. *Archaeopteryx,* the first bird, still possessed teeth, as did several fossils from the early history of birds. But no fossil bird has produced teeth during the past sixty million years, while the toothlessness of all modern birds ranks with wings and feathers as defining characters of the class. Nonetheless, although the system has not been used on its home ground for perhaps a hundred million generations, chick epithelium can still induce the formation of dentin when combined with appropriate mesenchyme (chick mesenchyme itself has probably lost the ability to form dentin, hence the toothlessness of hens and the necessity for using mice).

Kollar and Fisher then found something even more interesting. In four of their grafts, complete teeth had developed! Chick epithelium had not only induced mouse mesenchyme to form dentin; it had also been able to generate enamel matrix proteins. (Dentin must be induced by epithelium, but this epithelium cannot differentiate into enamel unless it, in turn, can interact with the very dentin it has induced. Since chick mesenchyme cannot form dentin, chick epithelium never gets the chance to show its persistent stuff in nature.)

One final point stunned me even more. Kollar and Fisher write of their best tooth: "The entire tooth structure was well formed, with root development in proper relation to the crown, but the latter did not have the typical first-molar morphology, since it lacked the cusp pattern usually present in intraocular grafts of first-molar rudiments." In other words, the tooth looks normal, but it does not have the form of a mouse's molar. The odd form may, of course, simply result from the peculiar interaction of two systems not meant to be joined in nature. But is it possible that we are seeing, in part, the actual form of a latent bird's tooth—the potential structure that chick epithelium has encoded for sixty million years but has not expressed in the absence of dentin to induce it?

Kollar and Fisher's work recalled another experiment from the opposite end of a chick, a famous story usually misreported by evolutionary biologists (once, I am embarrassed to say, by myself), as I discovered in tracking down the original source. In 1959, the French embryologist Armand Hampé reported some experiments on the development of leg bones in chick embryos. In ancestral reptiles, the tibia and fibula (the bones between your kneecap and ankle) are equal in length; the ankle region below includes a series of small bones. In *Archaeopteryx,* the first bird, tibia and fibula are still equal in length, but the ankle bones below have been reduced to two, one articulating with the tibia, the other with the fibula. In most modern birds, however, the fibula has been reduced to a splint. It never reaches the ankle region, while the two ankle bones are "engulfed" by the rapidly growing tibia and fuse with it. Thus, modern birds develop a single structure (the tibia with ankle bones fused to it and the rudimentary fibula at its side), articulating with bones of the foot below.

Hampé reasoned that the fibula might well maintain its capacity for attaining full, ancestral length, but that competition for material by the rapidly growing tibia might deprive it of any opportunity to express this potential. He therefore performed three types of experiments, all directed toward giving the fibula some relief from its imperialistic and normally victorious neighboring bone. In all cases, the fibula attained its ancestral length, equal to the tibia and reaching the ankle region below. In the first, Hampé simply grafted more embryonic tissue into the region of the growing leg bones. The tibia reached its characteristic length, but the region now had enough material "left over" for the fibula. In the second, he altered the direction of growth for tibia and fibula so that the two bones did not remain in intimate contact. In the third, he inserted a mica plate between the two bones; the developing tibia could no longer "grab" material from its less vigorous neighbor and the fibula achieved its full length.

Thus, Hampé recreated an ancestral relationship between two bones by a series of simple manipulations. And this alteration engendered an even more interesting consequence. In normal chicks, the fibula begins its growth in contact with one of the small ankle bones below. But as the tibia enlarges and predominates, this contact breaks at about the fifth day of development. The fibula then retreats to form its splint, while the expanding tibia engulfs both ankle bones to form a single structure. In one case during Hampé's manipulations, the two ankle bones remained separate and did not fuse with either tibia or fibula (while both ankle bones fused with the tibia, as usual, in the other leg of the same embryo—an untreated control allowed to develop normally). In this bird, Hampé's simple manipulation not only produced its intended result (expression of an ancestral relationship in leg bones); it also evoked the ancestral pattern of ankle bones as well.

Hampé was able to produce these impressive atavisms by simple manipulations that amount to minor, quantitative changes in timing of development or placement of embryonic tissue. Adding more tissue doesn't simply make a bigger part with the same proportions; it leads to differential growth of one bone (the fibula) and a change in arrangement of the entire ankle area (two ankle bones, articulating separately to tibia and fibula in some cases, rather than a single tibia with both ankle bones fused to it).

Developmental patterns of an organism's past persist in latent form. Chicks no longer develop teeth because their own mesenchyme does not form dentin, even though their epithelium can still produce enamel and induce dentin in other animals. Chicks no longer develop separate ankle bones because their fibula no longer keeps pace with the tibia during growth, but the ankle bones develop and retain their identity when fibulas are coaxed to reach their ancestral length. An organism's past not only constrains its future; it also provides as legacy an enormous reservoir of potential for rapid morphological change based upon small genetic alterations.

Charles Darwin constructed his theory as a two-stage process: variation to supply raw material and natural selection to impart direction. It is frequently (and incorrectly) stated that he said little about variation, embarrassed as he was by ignorance about the mechanism of heredity. Many people believe that he simply treated variation as a "black box," something to be assumed, mentioned in passing, and then forgotten. After all, if there is always enough variation for natural selection to use, why worry about its nature and causes?

Yet Darwin was obsessed with variation. His books, considered as an ensemble, devote much more attention to variation than to natural selection, for he knew that no satisfactory theory of major evolutionary change could be constructed until the causes of variation and the empirical rules of its form and amount had been elucidated. His longest book is devoted entirely to problems of variation—the two-volume *Variation of Animals and Plants Under Domestication* (1868). Darwin felt that atavism held the key to many mysteries of variation, and he devoted an entire chapter to it, closing (as I will) with these words:

> The fertilized germ of one of the higher animals . . . is perhaps the most wonderful object in nature. . . . On the doctrine of reversion [atavism] . . . the germ becomes a far more marvellous object, for, besides the visible changes which it undergoes, we must believe that it is crowded with invisible characters . . . separated by hundreds or even thousands of generations from the present time: and these characters, like those written on paper with invisible ink, lie ready to be evolved whenever the organization is disturbed by certain known or unknown conditions.

15 | Helpful Monsters

MY GRANDFATHER, who taught me to play poker and watched the Friday night fights with me every week, once took me to one of the cruelest, yet most fascinating spectacles of decades now thankfully past—the rows of malformed people forced (by an absence of other opportunities) to display themselves to a gawking public at the Ringling Brothers sideshow.

The genteel and legitimate counterpart to such public cruelty is the vast scientific literature on deformed births—a subject dignified with its own formal name as teratology, literally, the study of monsters. Although scientists are as subject as all people to the mixture of awe, horror, and curiosity that draws people to sideshows, teratology has an important rationale beyond primal fascination.

The laws of normal growth are best formulated and understood when the causes of their exceptions can be established. The experimental method itself, a touchstone of scientific procedure, rests upon the notion that induced and controlled departures from the ordinary can lay bare the laws of order. Congenitally malformed bodies are nature's experiments, uncontrolled by intentional human art to be sure, but sources of insight nonetheless.

The early teratologists sought to understand malformations by classifying them. In the decades before Darwin, French medical anatomists developed three categories: missing parts *(monstres par défaut),* extra parts *(monstres par excès),* and normal parts in the wrong places. The folklore of monsters had long recognized the last category in tales of anthropophagi, maneaters with eyes in their shoulders and a mouth on their breast. Shakespeare alluded both to them and to some related colleagues in *Othello* when he spoke of "The Anthropophagi and men whose heads/Do grow beneath their shoulders."

But a classification is no more than a set of convenient pigeonholes until the causes of ordering can be specified. And here nineteenth-century teratology got becalmed in its own ignorance of heredity. The establishment of genetics in our century revived a waning interest in teratology, as early Mendelians discovered the mutational basis of several common deformities.

Geneticists had particular success with one common category in the old classification—normal parts in the wrong places. They studied their favorite animal, the fruit fly, *Drosophila melanogaster,* and found a variety of bizarre transpositions. In the first of two famous examples, the halteres (organs of balance) are transformed into wings, restoring to the aberrant fly its ancestral complement of four (normal flies, as members of the order Diptera, have two wings). In the second, legs or parts of legs replace a variety of structures in the head—antennae and parts of the mouth in particular. Mutations of this sort are called homeotic.

Not all misplacements of parts represent homeosis, and this restriction is a key to the evolutionary message I shall draw further on. William Bateson, who later invented the term genetics, defined as "homeotic" only those parts that replace an organ having the same developmental or evolutionary origin (the word comes from a Greek root for "similar"). Thus, halteres are the evolutionary descendants of wings, while insect antennae, mouthparts, and legs all differentiate from similar precursors in the embryonic segments, and all presumably evolved from an ancestor with a pair of simple and similar appendages on each adult body segment. We might refer to homeosis if a human developed a second pair of arms where his legs should be, but an extra pair of arms on the chest would not qualify.

Homeotic mutants are found on all four pairs of chromosomes in *D. melanogaster.* A 1976 review by W. J. Ouweneel includes a list that runs to three full pages. But the two most famous, best studied, and elaborate sets of homeotic mutations both reside on the right arm of the third chromosome.

The first set, called the bithorax complex and abbreviated BX-C, regulates the normal development and differentiation of the fly's posterior body segments. The larval fly is already divided into a series of segments, initially quite similar, that will differentiate into specialized adult structures. The first five larval segments build the adult head (the first forms anterior parts of the head; the second, the eyes and antennae; and the third through fifth, the various parts of the mouth). The next three segments, T_1, T_2, and T_3, form the thorax. Each will bear a pair of legs in the adult, building the normal insect complement of six. The single pair of wings will differentiate in T_2.

The next eight segments (A_1 through A_8) form the adult's abdomen, while the final, or caudal, segment (A_9 and A_{10}) will build the adult's posterior end. The presence of normal BX-C genes appears to be a precondition for the ordinary development of all segments behind the second thoracic. If all the genes of BX-C are deleted from the third chromosome, all larval segments behind the second thoracic (T_2) fail to differentiate along their normal route and seem to become second thoracic segments themselves. If the adult survived, it would be a wonder to behold, with (presumably) a pair of legs on each of its numerous posterior segments. But this deletion is, in geneticist's jargon, "lethal," and the fly dies while still a larva. We know that the posterior segments of such aberrant flies are slated to develop as second thoracics because incipient differentiation within the larval segments serves as a sure guide to their later fate.

The bithorax complex includes at least eight genes, all located in sequence right next to each other. Edward B. Lewis (see bibliography), the distinguished geneticist from CalTech who has spent twenty years probing the complexities of BX-C, believes that these eight genes arose as repetitions of a single ancestral gene and then evolved in different directions. Just as the entire deletion of BX-C produces the striking homeotic effect of converting all posterior segments to second thoracics, several mutations in the eight genes produce homeotic results as well. The most famous mutation, called bithorax and commandeered as a name for the entire complex, converts the third thoracic segment into a second thoracic. Thus, the adult fly develops with two second thoracics and two pairs of wings, instead of one pair and a pair of halteres behind. (It is misleading to state that halteres "turn into" wings. Rather, the entire segment normally destined to be a third thoracic, and to produce halteres, develops as a second thoracic and builds wings.) In another mutation, called bithoraxoid, the first abdominal segment develops as a third thoracic, builds a pair of legs, and produces a fly with more than the usual insect number of six.

Lewis has proposed an interesting hypothesis for the normal action of BX-C genes. He believes that they are initially repressed (turned off) in the larval fly. As the fly develops, BX-C genes are progressively derepressed (turned on). The BX-C genes act as regulators—that is, they do not build parts of the body themselves but are responsible for turning on the structural genes that do code for building blocks. Adult form reflects the amount of BX-C gene-product in an embryonic segment; the more BX-C, the more posterior in appearance the segment. Lewis then argues that BX-C genes are derepressed in sequence, from the anterior point of their action (the third thoracic segment) to the back end of the animal. When a BX-C gene turns on, its product accumulates in a given segment and, simultaneously, in all segments posterior to it. BX-C first turns on in the third thoracic, and its product accumulates in all segments from the third thoracic to the posterior end. The next BX-C gene turns on in the next posterior segment, the first abdominal, and its product accumulates in all segments from the first abdominal to the posterior end. The next gene turns on in the second abdominal, and so forth. Thus, a gradient of BX-C product forms, with lowest concentration in the second thoracic and increasing amounts in a posterior direction. The more gene product, the more posterior in appearance the form of a resultant segment.

This hypothesis is consistent with the known homeotic effects of BX-C mutations. If BX-C is deleted entirely, it supplies no gene product, and all segments behind the second thoracic differentiate as second thoracics. In a mutation with opposite effect, all the BX-C genes are turned on at the same time in all segments—and all segments affected by BX-C then differentiate as eighth abdominals.

The second outstanding set of homeotics is also named for its most famous mutation—the antennapedia complex, or ANT-C. The fine structure of this complex has recently been elucidated in a series of remarkable experiments by Thomas C. Kaufman, Ricki Lewis, Barbara Wakimoto, and Tulle Hazelrigg in Kaufman's laboratory at the University of Indiana. (I thank Dr. Kaufman for introducing me to the literature of homeosis and for patient and lucid explanations of his own work.) The BX-C genes regulate the morphology of segmentation from the third thoracic to the posterior end; ANT-C also affects the third thoracic, but then regulates development in the five segments anterior to it (the other two thoracics and the three that produce parts of the mouth). If the entire complex is deleted, then all three thoracic segments begin to differentiate as first thoracics (while the abdominals, regulated by BX-C, develop normally). Apparently, the genes of ANT-C normally turn on in the second thoracic segment and trigger the proper development of the second and third thoracics.

Kaufman and his colleagues have found that ANT-C consists of at least seven genes, not all with known homeotic effects, lying right next to each other on the right arm of the third chromosome in *D. melanogaster.* The genes are not named for their normal effects (which, after all, just yield an ordinary fly bearing nothing special for recognition) but for their rare homeotic mutations. The first, the antennapedia gene, regulates differentiation of the second thoracic segment, and normally turns on there to accomplish its appointed function. A series of mutations has been detected at this locus, all with homeotic effects consistent with this interpretation of normal function.

One dominant mutation, antennapedia itself, has the bizarre effect (as its name implies) of producing a leg where an antenna ought to be. This wayward appendage is not any old leg, but clearly a second thoracic. The antennapedia mutation apparently works by turning on in the wrong place —the antennal segment—rather than in the second thoracic segment.

Another dominant mutation, called extra sex combs, leads to the appearance of sex combs on all three pairs of legs, not only on the first as in normal flies. This morphology is not simply the result of a gene that makes sex combs (contrary to popular belief, very few genes simply "make" individual parts without series of complex and coordinated effects). It is a homeotic mutation. All three thoracic segments differenti-

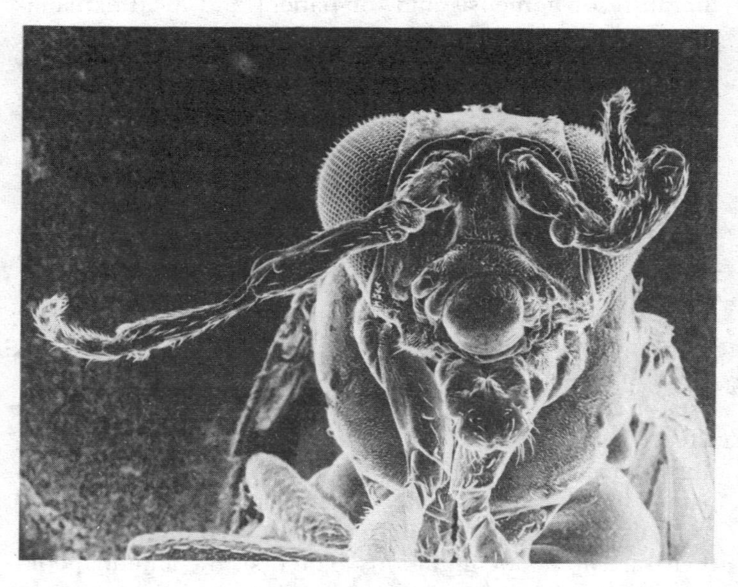

A fly with antennapedia mutation, in which legs form where the antenna should be. SCANNING ELECTRON MICROSCOPE PHOTO BY F. R. TURNER.

ate as first thoracics, and the fly has, literally, three pairs of first legs with their attendant sex combs. (The entire deletion of ANT-C also causes the three thoracics to differentiate as first thoracics, but this deletion is lethal and the fly dies in its larval stage. Flies with the extra sex comb mutation do live to become adults.) The extra sex comb mutation probably operates by suppressing the normal action of its gene. Since normal action causes the second thoracic segment to differentiate properly, suppression induces all three thoracics to develop as first thoracics.

The second gene of ANT-C is named for its prominent mutation, reduced sex comb. This homeotic mutation, contrary to the effect of its neighbor extra sex comb, causes the first thoracic segment to differentiate as a second thoracic. Only first thoracic legs bear sex combs in *D. melanogaster.* The next three genes do not have known homeotic effects, and their inclusion within ANT-C is something of a puzzle. One deletes a number of embryonic segments in its most prominent mutation, another interferes with normal development of the maxilla and mandible of the mouth, while the third produces a curiously wrinkled embryo.

The sixth gene of ANT-C is named for the other famous homeotic mutant of this complex—proboscipedia, discovered in 1933 by two of the century's most famous geneticists, Calvin Bridges and Theodosius Dobzhansky. Six mutations have been detected at this locus; most, like proboscipedia itself, produce legs where parts of the mouth should develop. The seventh and last (known) gene of ANT-C lacks homeotic effects in its mutant form and produces severe constrictions at segment boundaries in the larva. It is lethal.

Homeosis is not peculiar to fruit flies, but seems to be a general phenomenon, at least in arthropods. A set of mutations analogous (or even homologous) with the bithorax homeotics of *Drosophila* occurs in the silk moth *Bombyx* (order Lepidoptera; flies belong to the order Diptera). Two species of *Tribolium,* the flour beetle (order Coleoptera), exhibit mutations with effects that mimic the ANT-C homeotics of *Drosophila.* One set, in *Tribolium castaneum,* acts like antennapedia and produces a graded series of partial replacements of antennae by legs, ranging from tarsal claws on the eleventh antennal segment to the virtual replacement of an entire antenna with a foreleg. Another, in *T. confusum,* acts like proboscipedia and substitutes legs of the first thoracic segment for mouth structures known as labial palps. In the cockroach *Blatella germanica,* a homeotic mutant produces rudimentary wings on the first thoracic segment. No modern insect normally bears wings on its first thoracic segment, but the earliest winged fossil insects did!

Homeosis is easiest to demonstrate in arthropods with their characteristic body plan of discrete segments with different and definite fates in normal development, but common embryological and evolutionary origins. Yet analogous phenomena have been noted again and again in other animals and plants with repeated parts. In fact, Bateson's first example after defining the term cited vertebrae in the human backbone. All mammals (except sloths, but including giraffes) have seven cervical, or neck, vertebrae (they are awfully large in giraffes). These are followed by dorsal, or rib-bearing, vertebrae. Bateson noted numerous cases of humans with ribs on the seventh, and even a few with ribs on the sixth, cervical vertebra.

Homeotic mutants are gripping in their weirdness, but what do they teach us about evolution? We must avoid, I believe, the tempting but painfully naïve idea that they represent the long-sought "hopeful monsters" that might validate extreme saltationist views of major evolutionary transitions in single steps (a notion that I, despite my predilections for rapid change, regard as a fantasy born of insufficient appreciation for organisms as complex and integrated entities). First of all, most homeotic mutations produce hopeless creatures. The legs that extend from antennal sockets or surround mouths in afflicted flies are useless appendages without proper neural and muscular hookups. Even if they did work, what could they accomplish in such odd positions? Secondly, the viable homeotics mimicking ancestral forms are not really forebears reborn. A bithorax fly bears the ancestral complement of four wings, but it attains this state by growing two second thoraxes, not by recovering an ancient pattern.

I believe that the lessons of homeosis lie first in embryology and then cycle back to evolution. As Tom Kaufman pointed out to me, they demonstrate in a dramatic way how few genes are responsible for regulating the basic order of developing parts in a fruit fly's body. Together, the ANT-C and BX-C complexes of *D. melanogaster* specify the normal development of all the mouth, thoracic, and abdominal segments—only the two anterior segments are not subject to their control. Each complex contains only a handful of genes and each handful may have evolved from a single ancestral gene that repeated itself several times. When these genes mutate or are deleted, peculiar homeotic effects arise that usually throw development awry and lead to death.

Most importantly perhaps, these homeotic complexes display the hierarchical way in which genetic programs regulate the immense complexity of embryonic development, recognized since Aristotle's time as biology's greatest mystery. The homeotic genes do not build the different structures of each body segment themselves. This is the role of so-called structural genes that direct the assembly of proteins. The homeotics are switches or regulators; they produce some signal (of utterly unknown nature) that turns on whole blocks of structural genes.

Yet, at a higher level, some master regulator must be responsible for turning on the homeotics at the right time and in the right place, for we know that many homeotic mutations are mistakes in placement and timing. Perhaps this master regulator is no more than a gradient of some substance running from the front to the back end of a larval fly; perhaps the homeotic regulators can "read" this gradient and turn on in the right place by assessing its concentration. In any case, we have three hierarchical levels of control: the structural genes that build different parts in each segment, the homeotic regulators that switch on the blocks of structural genes, and the higher regulators that turn on the homeotic regulators in the right place and at the right time.

If embryology is a hierarchical system with surprisingly few master switches at high levels, then we might draw an evolutionary message after all. If genetic programs were beanbags of independent genes, each responsible for building a single part of the body, then evolution would have to proceed bit by bit, and any major change would have to occur slowly and sequentially as thousands of parts achieved their independent modifications. But genetic programs are hierarchies with master switches, and small genetic changes that happen to affect the switches might engender cascading effects throughout the body. Homeotic mutants teach us that small genetic changes can affect the switches and produce remarkable changes in an adult fly. Major evolutionary transitions may be instigated (although not finished all at once as hopeful monster enthusiasts argue) by small genetic changes that translate into fundamentally altered bodies. If classical Darwinian gradualism is now under attack in evolutionary circles, the hierarchical structure of genetic programs forms a powerful argument for the critics.

In this context, we consider the hypothetical major steps in insect evolution and recognize that homeotic mutants may help to illuminate them. Insects, with their relatively few, differentiated segments, probably evolved from an ancestor with more numerous and less differentiated segments. Initially, these less differentiated segments each bore a pair of legs (the antennae and mouthparts of modern insects are modified legs). Insects evolved by suppressing legs on the posterior segments and modifying them to antennae and mouthparts on the anterior segments. The major homeotic complexes of *Drosophila* seem to regulate just these changes—and with a minimum of genetic information. BX-C controls the posterior appendages with their suppressed legs, and its deletion causes these segments to begin a differentiation as second thoracics with incipient legs. The major mutants of ANT-C replace structures that were once legs with legs. The nature of homeotic changes is not capricious, but follows evolutionary channels.

Even the bizarre homeotics may not be devoid of evolutionary information. When Bridges and Dobzhansky described proboscipedia in 1933, they noted that a large set of coordinated changes—quite apart from the spectacular appearance of legs—all brought the mouthparts closer to the standard form of biting insects from which flies presumably evolved. (Dobzhansky, who died just a few years ago, was the greatest evolutionary geneticist of our times. Fancy, quantitative lab work often wins all the kudos while field naturalists, with their detailed and specific knowledge, are unfairly dismissed as stamp collectors. Dobzhansky's life proves how misguided this prejudice is. Geneticists had been describing homeotic mutants for years, but none had the knowledge to recognize the subtle morphological effects that require a trained taxonomist's eye to comprehend. Dobzhansky, the finest geneticist of them all, was a trained taxonomist and field biologist who began his work by specializing on the systematics of the Coccinellidae, or lady beetles. There is no substitute for detailed knowledge of natural history and taxonomy.)

If anyone has wondered whether homeotic mutants must find their significance only in highfalutin realms of evolutionary speculation, I close with an arresting fact. A homeotic mutation has been found in the biting mosquito *Aedes albopictus.* Yes, you guessed it. This mutation converts part of the biting apparatus into a pair of legs! The six stylets that actually pierce our skin are unaffected, but the labella, the structures that surround the stylets and contain tactile and chemosensory hairs, are converted to legs with tarsal claws at their tips. These mosquitos cannot pierce skin, both because they lack the tactile and chemosensory hairs that locate the right spot and because the stylets get entrapped in the misplaced legs.

What a wonderful and joyous idea in a world inundated with bad news—an ouchless mosquito with an extra pair of legs. Oh, don't raise your hopes. They won't replace the normal ones. First of all, they die because they cannot feed (although they can be artificially maintained on blood-soaked cotton balls). Even if they learned to feed by lapping instead of piercing, they would be no match for the normal kind because they have longer larval lives, increased pupal mortality, and a significant decrease in adult longevity. Still, these are the curious facts that nurture hope in parlous times—in this case, and with only a little poetic license, an enormous advantage (if only for another long-suffering creature) of putting a foot in one's mouth.

4 | Teilhard and Piltdown

16 The Piltdown Conspiracy[1]

Introduction and Background

OF CONSPIRACIES

IN HIS GREAT ARIA "La calunnia," Don Basilio, the music master of Rossini's *Barber of Seville,* graphically describes how evil whispers grow, with appropriate watering, into truly grand and injurious calumnies. For the less conniving among us, the same lesson may be read with opposite intent: in adversity, try to contain. The desire to pin evil deeds upon a single soul acting alone reflects this strategy; conspiracy theories have a terrible tendency to ramify like Basilio's whispers until the runaway solution to "whodunit" becomes "everybodydunit." But conspiracies do occur. Even the pros and pols now doubt that Lee Harvey Oswald

[1]This essay has been the subject of many commentaries, most positive, but some (to put it mildly) excoriatingly negative (the most brutal, if you'll pardon the aroma of suspicion by association, from devotees of the "Teilhard cult"). In this light, I have decided to reprint the essay (which first appeared in *Natural History Magazine* for August, 1980) without any changes—for it would be unfair to improve it by correcting errors and ambiguities and then to turn on my detractors (see next essay) with a product better than the original that they first criticized. Since it is simply immoral to publish known error, I shall correct a few mistakes by footnote, so everyone can see where I goofed the first time round. I shall let all interpretations stand without comment and indicate any changes of opinion (minor and insubstantial) in the next essay.

acted alone; and everybody did do it on the Orient Express.[2]

The Piltdown case, surely the most famous and spectacular fraud of twentieth-century science, has experienced this tension ever since its exposé in 1953. The semiofficial, contained version holds that Charles Dawson, the lawyer and amateur archeologist who "found" the first specimens, devised and executed the entire plot himself. Since J. S. Weiner's elegant case virtually precludes Dawson's innocence (*The Piltdown Forgery,* Oxford University Press, 1955), conspiracies become the only reasonable refuge for challengers. And proposals for coconspirators abound, ranging from the great anatomist Grafton Elliot Smith to W. J. Sollas, professor of geology at Oxford. I regard these claims as farfetched and devoid of reasonable evidence. But I do believe that a conspiracy existed at Piltdown and that, for once, the most interesting hypothesis is actually true. I believe that a man who later became one of the world's most famous theologians, a cult figure for many years after his death in 1955, knew what Dawson was doing and probably helped in no small way—the French Jesuit priest and paleontologist Pierre Teilhard de Chardin.

TEILHARD AND PILTDOWN

Teilhard, born in Auvergne (central France) in 1881, belonged to an old, conservative, and prosperous family. Entering the Society of Jesus in 1902, he studied on the English island of Jersey from 1902 to 1905 and then spent three years as a teacher of physics and chemistry at a Jesuit school in Cairo. In 1908, he returned to finish his theological training at the Jesuit seminary of Ore Place in Hastings, providentially located right next to Piltdown on England's southeast coast. Here he stayed for four years, and here he

---

[2]I guess this should now read "Many pros and pols," the wheel of fortune spinning quickly, as it does. Yet fiction has, ironically, the rock-hard permanence that fact must lack—and everyone did do it, now and forever more, on the Orient Express.

was ordained a priest in 1912.[3] As a theological student, Teilhard was talented enough, but lackadaisical. His passion at Hastings was, as it always had been, natural history. He scoured the countryside for butterflies, birds, and fossils. And, in 1909, he met Charles Dawson at the focus of their common interests—in a stone quarry, hunting for fossils. The two men became good friends and colleagues in pursuit of their interest. Teilhard described Dawson to his parents as "my correspondent in geology."

Dawson claimed that he had recovered the first fragment of Piltdown's skull in 1908, after workmen at a gravel pit told him of a "coconut" (the entire skull) they had unearthed and smashed at the site. Dawson kept poking about, collecting a few more skull pieces and some fragments of other fossil mammals. He did not bring his specimens to Arthur Smith Woodward, keeper of paleontology at the British Museum, until the middle of 1912. Thus, for three years before any professional ever heard of the Piltdown material, Dawson and Teilhard were companions in natural history in the environs of Piltdown.

Smith Woodward was not a secretive man, but he knew the value of what Dawson had brought and the envy it might inspire. He clamped a tight lid upon Dawson's information prior to its publication. He wanted none of Dawson's lay friends at the site, and only one naturalist accompanied Dawson and Smith Woodward in their first joint excavations at Piltdown—Teilhard de Chardin, whom Dawson had described as "quite safe." More specimens came to light during 1912, including the famous jaw with its two molar teeth, artificially filed to simulate human patterns of wear. In December, Smith Woodward published and the controversy began.

The skull fragments, although remarkably thick, could not be distinguished from those of modern humans. The

---

[3] I thank Rev. Thomas M. King, S. J. of Georgetown University for pointing out two inconsequential but highly embarrassing errors in this paragraph. Teilhard entered the Socity of Jesus in 1899, not 1902, and he was ordained in 1911, not 1912.

jaw, on the other hand, except for the wear of its teeth, loudly said "chimpanzee" to many experts (in fact, it once belonged to an orangutan). No one smelled fraud, but many professionals felt that parts of two creatures had been mixed together at the Piltdown site. Smith Woodward stoutly defended the integrity of his creature, arguing, with flawed logic, that the crucial role of brain power in our mastery of the earth today implies a precocious role for large brains in evolutionary history as well. A fully vaulted skull still attached to an apish jaw vindicated such a brain-centered view of human evolution.

Teilhard left England late in 1912[4] to begin his graduate studies with Marcellin Boule, the greatest physical anthropologist of France. But in August 1913, he was back in England for a retreat at Ore Place. He also spent several days prospecting with Dawson and on August 30 made a major discovery himself—a canine tooth of the lower jaw, apish in appearance but worn in a human fashion. Smith Woodward continued his series of publications on the new material, but critics persisted in their belief that Piltdown man represented two animals improperly united.

The impasse broke in Smith Woodward's favor in 1915. Dawson had been prospecting at another site, two miles from Piltdown, for several years. He probably took Teilhard there in 1913; we know that he searched the area several times with Smith Woodward in 1914. Then, in January 1915, he wrote to Smith Woodward. The second site, later called Piltdown 2, had yielded its reward: "I believe we are in luck again. I have got a fragment of the left side [it was actually the right] of a frontal bone with a portion of the orbit and root of nose." In July of the same year, he announced the discovery of a lower molar, again, apish in

[4]Again I thank Rev. King for calling my attention to an error, this one of some potential consequence and therefore more embarrassing. The longer Teilhard remained in England, the more opportunity he had to work with Dawson. He left, in fact, not "late in 1912," as I stated, but on July 16.

appearance but worn in a human fashion. The bones of a human and an ape might wash into the same gravel pit once, but the second, identical association of vaulted skull and apish jaw surely proved the integrity of a single bearer, despite the apparent anatomical incongruity. H. F. Osborn, America's leading paleontologist and critic of the first Piltdown find, announced a conversion in his usual grandiloquent fashion. Even Teilhard's teacher Marcellin Boule, leader of the doubters, grumbled that the new finds had tipped the balance, albeit slightly, in Smith Woodward's favor. Dawson did not live to enjoy his triumph, for he died in 1916. Smith Woodward stoutly supported Piltdown for the rest of his long life, devoting his last book (*The Earliest Englishman,* 1948) to its defense. He died, mercifully, before his bubble burst.

Meanwhile, Teilhard pursued his calling with mounting fame, frustration, and exhilaration. He served with distinction as a stretcher bearer in World War I and then became professor of geology at the Institut Catholique of Paris. But his unorthodox (although always pious) thinking soon led him into irrevocable conflict with ecclesiastical authority. Ordered to abandon his teaching post and to leave France, Teilhard departed for China in 1926. There he remained for most of his life, pursuing distinguished research in geology and paleontology and writing the philosophical treatises on cosmic history and the reconciliation of science with religion that later made him so famous. (They all remained unpublished, by ecclesiastical fiat, until his death.) Teilhard died in 1955, but his passing only marked the beginning of his meteoric rise to fame. His treatises, long suppressed, were published and quickly translated into all major languages. *The Phenomenon of Man* became a best seller throughout the world. Harvard's Widener Library now houses an entire tier of books devoted to Teilhard's writing and thinking. Two journals that were established to discuss his ideas still flourish.

Of the original trio—Dawson, Teilhard, and Smith Woodward—only Teilhard was still living when Kenneth

Oakley, J. S. Weiner, and W. E. le Gros Clark proved that the Piltdown bones had been chemically stained to mimic great age, the teeth artificially filed to simulate human wear, the associated mammal remains all brought in from elsewhere, and the flint "implements" recently carved. The critics had been right all along, more right than they had dared to imagine. The skull bones did belong to a modern human, the jaw to an orangutan. As the shock of revelation gave way to the fascination of whodunit, suspicion quickly passed from two members of the trio. Smith Woodward had been too dedicated and too gullible; moreover, he knew nothing of the site before Dawson brought him the original bones in 1912. (I have no doubt whatsoever of Smith Woodward's total innocence.) Teilhard was too famous and too present for any but the most discreet probing. He was dismissed as a naïve young student who, forty years before, had been duped and used by the crafty Dawson. Dawson acting alone became the official theory; professional science was embarrassed, but absolved.

DOUBTS

I was just the right age for primal fascination—twelve years old—and a budding paleontologist when news of the fraud appeared on page one of the *New York Times* one morning at breakfast. My interest has never abated, and I have, over the years, asked many senior paleontologists about Piltdown. I have also remarked, both with amusement and wonder, that very few believed the official tale of Dawson acting alone. I noted, in particular, that several of the men I most admire suspected Teilhard, not so much on the basis of hard evidence (for their suspicions rested on what I regard as a weak point among the arguments), but from an intuitive feeling about this man whom they knew well, loved, and respected, but who seemed to hide passion, mystery, and good humor behind a garb of piety. A. S. Romer and Bryan Patterson, two of America's leading vertebrate paleontologists and my former colleagues at Harvard, often voiced their suspicions to me. Louis Leakey voiced them in print, without naming the name, but with no ambiguity for anyone in the know (see his autobiography, *By the Evidence*).[5]

I finally decided to get off my butt and probe a bit after I wrote a column on Piltdown for other reasons (*Natural History,* March 1979). I read all the official documents and concluded that nothing excluded Teilhard, although nothing beyond his presence at Piltdown from the start particularly implicated him either. I intended to drop the subject or to pass it along to someone with a greater zeal for investigative reporting. But at a conference in France last September, I happened to meet two of Teilhard's closest colleagues, the leading paleontologist J. Piveteau and the great zoologist P. P. Grassé. They greeted my suspicions with a blustering "*incroyable.*" Then Père François Russo, Teilhard's friend and fellow Jesuit, heard of my inquiries and promised to send me a document that would prove Teilhard's innocence—a copy of the letter that Teilhard had written to Kenneth Oakley on November 28, 1953. I received this letter in printed French translation (Teilhard wrote it in English) in October 1979 and realized immediately that it contained an inconsistency (a slip on Teilhard's part) most easily resolved by the hypothesis of Teilhard's complicity. When I visited Oakley at Oxford in April 1980, he showed me the original letter along with several others that Teilhard had written to him. We studied the documents and discussed Piltdown for the better part of a day, and I left convinced that Romer, Patterson, and Leakey had been right. Oakley, who had noted the inconsistency but interpreted it differently, agreed with me and stated as we parted: "I think it's right that Teilhard was in it." (Let me here express my deep appreciation for Dr. Oakley's hospitality, his openness, and his simple, seemingly inexhaustible kindness and helpfulness. I always feel so exhilarated when I discover—and it is not so rare as many people imagine—that a great thinker is also an exemplary human being.)

---

[5]I have since learned that Louis Leakey was far more serious in his probings than I had realized. He was convinced of Teilhard's guilt and was writing a book on the subject when he died.

Since then, I have sharpened the basic arguments and read through Teilhard's published work, finding a pattern that seems hard to reconcile with his innocence. My case is, to be sure, circumstantial (as is the case against Dawson or anyone else), but I believe that the burden of proof must now rest with those who would hold Father Teilhard blameless.

The Case against Teilhard

THE LETTERS TO KENNETH OAKLEY

The main virtue of truth, quite apart from its ethical value (which I hold to be considerable), is that it represents an infallible guide for keeping your story straight. The problem with prevarication is that, when the going gets complex or the recollection misty, it becomes very difficult to remember all the details of your invented scheme. Richard Nixon finally succumbed on a minor matter, and Sir Walter Scott spoke truly when he wrote the famous couplet: "Oh, what a tangled web we weave,/When first we practice to deceive!"

Teilhard made just such a significant slip on a minor point in his letter to Oakley. Teilhard offered no spontaneous recollections about Piltdown and responded only to Oakley's direct inquiries for help in establishing the forger's identity. He begins by congratulating Oakley "most sincerely on your solution of the Piltdown problem. Anatomically speaking, *'Eoanthropus'* [Smith Woodward's name for the Piltdown animal] was a kind of monster. . . . Therefore I am fundamentally pleased by your conclusions, in spite of the fact that, sentimentally speaking, it spoils one of my brightest and earliest paleontological memories."

Teilhard then stonewalls on the question of fraud. He refuses to believe it at all, declaring that Smith Woodward and Dawson (and, by implication, himself) were not the kind of men who could conceivably do such a thing. Is it not possible, he asks, that some collector discarded the ape bones in a gravel pit that legitimately contained a human skull, the product of a recent interment? Could not the iron staining have been natural, since the local water "can stain (with iron) at a remarkable speed"? But Teilhard's notion can explain neither the artificial filing of the teeth to simulate human wear nor the crucial discovery of a second combination of ape and human at the Piltdown 2 site. In fact, Teilhard admits: "The idea sounds fantastic. But, in my opinion, no more fantastic than to make Dawson the perpetrator of a hoax."

Teilhard then goes on to discuss Piltdown 2 and, in trying to exonerate Dawson, makes his fatal error. He writes:

> He [Dawson] just brought me to the site of Locality 2 and explained me [*sic*] that he had found the isolated molar and the small pieces of skull in the heaps of rubble and pebbles raked at the surface of the field.

But this cannot be. Teilhard did visit the second site with Dawson in 1913, but they did not find anything. Dawson "discovered" the skull bones at Piltdown 2 in January 1915, and the tooth not until July 1915. And now, the key point: Teilhard was mustered into the French army in December 1914 and was shipped immediately to the front, where he remained until the war ended. He could not have seen the remains of Piltdown 2 with Dawson, unless they had manufactured them together before he left (Dawson died in 1916).

Oakley caught the inconsistency immediately when he received Teilhard's letter in 1953, but he read it differently and for good reason. At that time, Oakley and his colleagues were just beginning their explorations into whodunit. They rightly suspected Dawson and had written to Teilhard to gather evidence. Oakley read Teilhard's statement when he was simply trying to establish the basic fact of Dawson's guilt. In that context, he assumed that Dawson had shown the specimens to Teilhard in 1913, but had withheld them from

Smith Woodward until 1915—more evidence for Dawson's complicity.

Oakley wrote back immediately, and Teilhard, realizing that he had tripped, began to temporize. In his second letter of January 29, 1954, he tried to recoup:

> Concerning the point of "history" you ask me, my "souvenirs" are a little vague. Yet, by elimination (and since Dawson died *during* the First World War, if I am correct) my visit with Dawson to the *second* site (where the two small fragments of skull and the isolated molar were found in the rubble) must have been in late July 1913 [it was probably in early August].

Obviously troubled, he then penned the following postscript.

> When I visited the site no. 2 (in 1913?) the two small fragments of skull and tooth had already been found, I believe. But your very question makes me doubtful! Yes, I think definitely they *had* been already found: and that is the reason why Dawson pointed to me the little heaps of raked pebbles as the place of the "discovery."

In a final letter to Mable Kenward, daughter of the owner of Barkham Manor, site of the first Piltdown find, Teilhard drew back even further: "Dawson showed me the field where the second skull (fragments) were found. But, as I wrote to Oakley, I cannot remember whether it was after or before the find" (March 2, 1954).

I can devise only four interpretations for Teilhard's slip.

1. I thought initially, when I had only read the first letter, that one might interpret Teilhard's statement thus: Dawson took me to the site in 1913 and later stated in wartime correspondence that he had found the fragments in the rubble. But Teilhard's second letter states explicitly that Dawson, in the flesh, had pointed to the spot at Piltdown 2 where he had found the specimens.

2. Oakley's original hypothesis: Dawson showed the specimens to an innocent Teilhard in 1913, but withheld them from Smith Woodward until 1915. But Dawson would not blow his cover in such a crude way. For Dawson took Smith Woodward to the second site on several prospecting trips in 1914, always finding nothing. Now Teilhard and Smith Woodward were also fairly close. Dawson had introduced them in 1909 by sending to London some important mammal specimens (having nothing to do with Piltdown) that Teilhard had collected. Smith Woodward was delighted with Teilhard's work and praised him lavishly in a publication. He accepted Teilhard as the only other member of their initial collecting trips at Piltdown. Moreover, Teilhard was a house guest of the Smith Woodwards when he visited London in September 1913, following his discovery of the canine. If Dawson had shown Teilhard the Piltdown 2 finds in 1913, then led Smith Woodward extensively astray during several field trips in 1914, and if an innocent Teilhard had told Smith Woodward about the specimens (and I can't imagine why he would have held back), then Dawson would have been exposed.

3. Teilhard never did hear about the Piltdown 2 specimens from Dawson, but simply forgot forty years later that he had never actually viewed the fossils he had read about later. This is the only alternative (to Teilhard's complicity) that I view as at all plausible. Were the letters not filled with other damaging points, and the case against Teilhard not supported on other grounds, I would take this possibility more seriously.

4. Teilhard and Dawson planned the Piltdown 2 discovery before Teilhard left England. Forty years later, Teilhard misconstructed the exact chronology, forgot that he could not have seen the specimens when they were officially "found," and slipped in writing to Oakley.

Teilhard's letters to Oakley contain other curious statements, each insignificant (or subject to other interpretations) by itself, but forming in their ensemble a subtle attempt to direct suspicion away from himself.

1. In his letter of November 28, 1953, Teilhard states that he first met Dawson in 1911. In fact, they met in May 1909, for Teilhard describes the encounter in a vivid letter to his parents. Moreover, this meeting was an important event in Teilhard's career, for Dawson befriended the young priest and personally forged his path to professional notice and respect by sending some important specimens he had collected to Smith Woodward. When Smith Woodward described this material before the Geological Society of London in 1911, Dawson, in the discussion following Smith Woodward's talk, paid tribute to the "patient and skilled assistance" given to him by Teilhard since 1909. I don't regard this, in itself, as a particularly damning point. A first meeting in 1911 would still be early enough for complicity (Dawson "found" his first piece of the Piltdown skull in 1911, although he states that a workman had given him a fragment "several years before"), and I would never hold a mistake of two years against a man who tried to remember the event forty years later. Still, the later (and incorrect) date, right upon the heels of Dawson's first "find," certainly averts suspicion from Teilhard.

2. Oakley wrote again in February 1954, probing further into Dawson's first contact with the Piltdown material, wondering in particular what had happened in 1908. Teilhard simply replied (March 1, 1954): "In 1908 I did not know Dawson." True enough, but they met just a few months later, and Teilhard might have mentioned it. A small point, to be sure.

3. In the same letter, Teilhard tries further to avert suspicion by writing of his years at Hastings: "You know, at that time, I was a young student in theology—not allowed to leave much his cell of Ore Place (Hastings)." But this description of a young, pious, and restricted man stands in stark contrast with the picture that Teilhard painted of himself at the time in a remarkable series of letters to his parents (*Lettres de Hastings et de Paris 1908–1912,* Paris: Aubier, 1965). These letters speak little of theology, but they are filled with charming and detailed accounts of Teilhard's frequent wanderings all over southern England. Eleven let-

ters refer to excursions with Dawson,[6] and no other naturalist is mentioned so frequently. If he spent much time at Ore Place, he didn't choose to write about it. On August 13, 1910, for example, he exclaims: "I have travelled up and down the coast, to the left and right of Hastings; thanks to the cheap trains [*les cheaptrains* as he writes in French] so common at this time of year, it is easy to go far with minimal expense."

Perhaps I am now too blinded by my own attraction to the hypothesis of Teilhard's complicity. Perhaps all these points are minor and unrelated, testifying only to the faulty memory of an aging man. But they do form an undeniable pattern. Still, I would not now come forward with my case were it not for a second argument, more circumstantial to be sure, but somehow more compelling in its persistent pattern of forty years—the record of Teilhard's letters and publications.

PILTDOWN IN TEILHARD'S WRITING

I remember a jokebook I had as a kid. The index listed "mule, sex life," but the indicated page was blank (ridiculous, in any case, for mules do not abstain just because the odd arrangement of their hybrid chromosomes debars them from bearing offspring). Teilhard's published record on Piltdown is almost equally blank. In 1920, he wrote one short article in French for a popular journal on *Le cas de l'homme de Piltdown.* After this, virtually all is silence. Piltdown never again received as much as a full sentence in all his published work (except once in a footnote). Teilhard mentioned Piltdown only when he could scarcely avoid it—in comprehensive review articles that discuss all outstanding human fossils. I can find fewer than half a dozen references in the twenty-three volumes of his complete works. In each case,

[6]Several critics have pointed out that some of the letters refer to visits that Dawson made to Teilhard's seminary at Ore Place rather than to field trips, or "excursions" proper. I goofed to be sure, but I don't see how my point is weakened.

Piltdown appears either as an item listed without comment in a footnote or as a point (also without comment) on a drawing of the human evolutionary tree or as a partial phrase within a sentence about Neanderthal man.[7]

Consider just how exceedingly curious this is. In his first letter to Oakley, Teilhard described his work at Piltdown as "one of my brightest and earliest paleontological memories." Why, then, such silence? Was Teilhard simply too

---

[7](This footnote, and only this one, was part of the original essay.) Teilhard's complete works are spread over two editions—a thirteen-volume edited compendium of his general articles (Paris: Editions de Seuil), and a more extensive ten-volume facsimile reprint of his professional publications (Olten: Walter-Verlag). To my annoyance, I discovered that the Paris edition has expurgated all Piltdown references without so stating and without even inserting ellipses. In trying to spiff up Teilhard's record, they have made him appear even more culpable by accentuating the impression of guilty silence. I therefore consulted likely originals whenever I could find them. One expurgation is particularly infuriating in its downright misrepresentation. A posthumous volume of essays, *Le Coeur de la matière* ("The heart of matter," Seuil, 1976), reprints Teilhard's application for the chair of paleontology at the Collège de France in 1948 (ecclesiastical authority did not permit him to accept). In this autobiographical essay, Teilhard discusses his role in human paleontology: "My first stroke of good luck in this area of ancient human paleontology came in 1923 when I was able to establish, with Emile Licent, the existence, hitherto contested, of paleolithic man in Northern China." If Teilhard had thus truly suppressed Piltdown in presenting himself for review, what but complicity could we infer? By sheer good fortune, I found a copy of this unpublished, mimeographed document in the reprint collection of my late colleague A. S. Romer. It reads: "My first stroke of good luck in this area of ancient human paleontology was to be included, when still young, in the excavation of *Eoanthropus dawsoni* in England. The second was, in 1923, to be able to establish, with Emile Licent. . . ." I doubt that half a dozen copies of the unpublished original exist in America, and I might easily have been led from the published and doctored version to an even stronger indication of Teilhard's complicity based on his silence. (It has, of course, occurred to me that the pattern of silence I detect in Teilhard's writing might represent more his editor's posthumous expurgation than Teilhard's own preference. Yet the completely honest ten-volume facsimile edition contains more than enough material to establish the pattern, and I have checked enough original versions of potentially expurgated texts to be confident that references to Piltdown are fleeting and exceedingly sparse throughout Teilhard's writing.)

diffident or saintly to toot his own trumpet? Scarcely, since no theme receives more voluminous attention, in scores of later articles, than his role in unearthing the legitimate Peking man in China.

As I began my investigation into this extraordinary silence, and trying to be as charitable as I could, I constructed two possible exonerating reasons for Teilhard's failure to discuss the major event of his paleontological youth. Kenneth Oakley then told me of the 1920 article, the only analysis of Piltdown that Teilhard ever published. I found a copy in the ten-volume edition of Teilhard's *oeuvre scientifique,* and realized that its content invalidated the only exculpatory arguments I could construct.

The first argument: Marcellin Boule, Teilhard's revered teacher, was a leading critic of Piltdown. He regarded it as a mixture of two creatures (not as a fraud), although he softened his opposition after he learned of the subsequent discovery at Piltdown 2. Perhaps Boule upbraided his young student for gullibility, and Teilhard, embarrassed to the quick, never spoke of the infernal creature, or of his role in discovering it, again.

The 1920 article invalidates such a conjecture, for in this work, Teilhard comes down squarely on the right side. He mentions that his English companions, convinced by finding the jaw so close to the skull fragments, never doubted the integrity of their fossil. Teilhard then notes, with keen insight, that experts who had not seen the specimens in situ would be swayed primarily by the formal anatomy of the bones themselves, and that these bones loudly proclaimed: human skull, ape's jaw. Which emphasis, then, shall prevail, geology or anatomy? Although he had witnessed the geology, Teilhard opted for anatomy:

> In order to admit such a combination of forms [a human skull and an ape's jaw in the same creature], it is necessary that we be forced to such a conclusion. Now this is not the case here. . . . The reasonable attitude is to grant primacy to the intrinsic morphological probability over the extrinsic probability of geologi-

> cal conditions. . . . We must suppose that the Piltdown skull and jaw belong to two different subjects.

Teilhard called it once, and he called it right. He had no reason to be embarrassed.

The second argument: Perhaps Teilhard had reveled in his role at Piltdown, cherished the memory, but simply found that the man he had helped to unearth could offer no support for, or even contact with, the concerns of his later career. On a broad level, this argument is implausible, if only because Teilhard wrote several general reviews about human fossils; however controversial or dubious, Piltdown should have been discussed. Even the leading doubters never failed to air their suspicions. Boule wrote chapters about Piltdown. Teilhard listed it without comment a few times and only when he had no choice.

In a more specific area, Teilhard's silence about Piltdown becomes inexplicable to the point of perversity (unless guilt and knowledge of fraud engendered it)—for Piltdown provided the best available support that fossils could provide for the most important argument of Teilhard's cosmic and mystical views about evolution, the dominant theme of his career and the source of his later fame. Teilhard never availed himself of his own best weapon, partly provided by his own hand.

The conclusion that skull and jaw belonged to different creatures did not destroy the scientific value of Piltdown, provided that both animals legitimately lay in the strata that supposedly entombed them. For these strata were older than any housing Neanderthal man, Europe's major claim to anthropological fame. Neanderthal, although now generally considered as a race of our species, was a low-vaulted, beetle-browed fellow of decidedly "primitive" cast. Piltdown, despite the thickness of its skull bones, looked more modern in its globular vault. The assignment of the jaw to a fossil ape further enhanced the skull's advanced status. Humans of modern aspect must have lived in England even before Neanderthal man evolved on the continent. Neanderthal, therefore, cannot be an ancestral form; it must represent a side branch of the human tree. Human evolution is not a ladder but a series of lineages evolving along separate paths.

In the 1920 article, Teilhard presented the Piltdown skull, divorced from its jaw, in just that light—as proof that hominids evolved as a bundle of lineages moving in similar directions. He wrote:

> Above all, it is henceforth proved that even at this time [of Piltdown] a race of men existed, already included in our present human line, and very different from those that would become Neanderthal. . . . Thanks to the discovery of Mr. Dawson, the human race appears to us even more distinctly, in these ancient times, as formed of strongly differentiated bundles, already quite far from their point of divergence. For anyone who has an idea of paleontological realities, this light, tenuous as it appears, illuminates great depths.

To what profundity, then, did Teilhard refer Piltdown as evidence? Teilhard believed that evolution moved in an intrinsic direction representing the increasing domination of spirit over matter. Under the thrall of matter, lineages would diverge to become more unlike, but all would move upward in the same general direction. With man, evolution reached its crux. Spirit had begun its domination over matter, adding a new layer of thought—the noosphere—above the older biosphere. Divergence would be stemmed; indeed, convergence had already begun in the process of human socialization. Convergence will continue as spirit prevails. When the last vestiges of matter have been discarded, spirit will involute upon itself at a single point called Omega and identified with God—the mystical evolutionary apocalypse that secured Teilhard's fame.[8]

But convergence is a thing of the future. Scientists seeking evidence for such a scheme must look to the past for

---

[8]Essay 18 presents a more detailed account of Teilhard's evolutionary philosophy.

twin signs of divergence accompanied by similar upward direction—in other words, for *multiple, parallel lineages* within larger groups.

I have read all of Teilhard's papers from the early 1920s. No theme receives more emphasis than the search for multiple, parallel lineages. In an article on fossil tarsiers, written in 1921, he argues that three separate primate lineages extend back to the dawn of the age of mammals, each evolving in the same direction of larger brains and smaller faces. In a review published in 1922 of Marcellin Boule's *Les hommes fossiles,* Teilhard writes: "Evolution is no more to be represented in a few simple strokes for us than for other living things; but it resolves itself into innumerable lines which diverge at such length that they appear parallel." In a general essay on evolution, printed in 1921, he speaks continually of oriented evolution in multiple, parallel lines within mammals.

But where was Piltdown in this extended paean of praise for multiple, parallel lineages? Piltdown provided proof, the only available proof, of multiple, parallel lineages within human evolution itself—for its skull belonged to an advanced human older than primitive Neanderthal. Piltdown was the most sublime argument that Teilhard possessed, and he never breathed it again after the 1920 article.

These two arguments have been abstract. A third feature of the 1920 article is stunning in its directness. For I believe that Teilhard fleetingly tried to tell his colleagues, too subtly perhaps, that Piltdown was a phony. In discussing whether the Piltdown remains represent one or two animals, Teilhard laments that the direct and infallible test cannot be applied. One skull fragment contained a perfect glenoid fossa, the point of articulation for the upper jaw upon the lower. Yet the corresponding point of the lower jaw, the condyle, was missing on a specimen otherwise beautifully preserved at its posterior end. Teilhard writes: "Since the glenoid fossa exists in perfect state on the temporal bone, we could simply have tried to articulate the pieces, *if the mandible had preserved its condyle:* we could have learned, without possible doubt, if the two fit together." I read this statement in a drowsy state one morning at two o'clock, but the next line—set off by Teilhard as a paragraph in itself terminated by an exclamation point—destroyed any immediate thought of sleep: "As if on purpose [*comme par exprès*], the condyle is missing!"

"*Comme par exprès.*" I couldn't get those words out of my mind for two days. Yes, it could be a literary line, a permissible metaphor for emphasis. But I think that Teilhard was trying to tell us something he didn't dare reveal directly.

OTHER ARGUMENTS

1. Teilhard's embarrassment at Oakley's disclosure. Kenneth Oakley told me that, although he had not implicated Teilhard in his thoughts, one aspect of Teilhard's reaction had always puzzled him. All other scientists, including those who had cause for the most profound embarrassment (like the aged Sir Arthur Keith, who had used Piltdown for forty years as the bedrock of his thought), expressed keen interest amidst their chagrin. They all congratulated Oakley spontaneously and thanked him for resolving an issue that had always been puzzling, even though the solution hurt so deeply. Teilhard said nothing. His congratulations arrived only when they could not be avoided—in the preface to a letter responding to Oakley's direct inquiries. When Teilhard visited London, Oakley tried to discuss Piltdown, but Teilhard always changed the subject. He took Teilhard to a special exhibit at the British Museum illustrating how the hoax had been uncovered. Teilhard glumly walked through as fast as he could, eyes averted, saying nothing. (A. S. Romer told me several years ago that he also tried to conduct Teilhard through the same exhibit, and with the same strange reaction.) Finally, Teilhard's secretary took Oakley aside and explained that Piltdown was a sensitive subject with Father Teilhard.

But why? If he had been gulled by Dawson at the site, he had certainly recouped his pride. Smith Woodward had devoted his life to Dawson's concoction. Teilhard had written about it but once, called it as correctly as he could, and then shut up. Why be so embarrassed? Unless, of course, the embarrassment arose from guilt about another aspect of his silence—his inability to come clean while he watched men he loved and respected make fools of themselves, partly on his account. Marcellin Boule, his beloved master, for example, correctly called Smith Woodward's *Eoanthropus* "an artificial and composite being" in the first edition of *Les hommes fossiles* (1921). The skull, he said, could belong to *"un bourgeois de Londres";* the jaw belonged to an ape. But he pondered the significance of Piltdown 2 and changed his mind in the second edition of 1923: "In the light of these new facts, I cannot be as sure as I was before. I recognize that the balance has now tipped a bit in the direction of Smith Woodward's hypothesis—and I am happy for this scientist whose knowledge and character I esteem equally." How did Teilhard feel as he watched his beloved master, Boule, falling into the abyss—when he contained tools for extraction that he could not use.

2. The elephant and the hippo. Bits and pieces of other fossil mammals were salted into the Piltdown gravels in order to set a geologic matrix for the human finds. All but two of these items could have been collected in England. But the hippo teeth, belonging to a distinctive dwarfed species, probably came from the Mediterranean island of Malta. The elephant tooth almost surely came from a distinctive spot at Ichkeul, Tunisia, for it is highly radioactive as a result of seepage from surrounding sediments rich in uranium oxide. This elephant species has been found in several other areas, but nowhere else in such highly radioactive sediments. Moreover, the Ichkeul site was only discovered by professionals in 1947; the doctored specimen at Piltdown could not have come from a cataloged museum collection.

Teilhard taught physics and chemistry at a Jesuit school in Cairo from 1905 to 1908, just before coming to Piltdown. His volume of *Letters from Egypt* again records little about theology and teaching, but much about travel, natural history, and collecting. He did not call at Tunisia or Malta on his passage down, but I can find no record of his passage back, and the two areas are right on his route from Cairo to France. In any case, Teilhard's letters from Cairo abound in tales of swapping and exchange with other natural historians of several North African nations. He was plugged into an amateur network of information and barter and might have received the teeth from a colleague.

This argument formed the base of evidence among my senior colleagues who suspected Teilhard—A. S. Romer, Bryan Patterson, and Louis Leakey (Leakey also mentioned Teilhard's knowledge of chemistry and the clever staining of the Piltdown bones). According to hearsay, le Gros Clark himself, a member of the trio that exposed the hoax, also suspected Teilhard on this basis. I regard this argument as suggestive, but not compelling. Dawson too was plugged into a network of amateur exchange.

3. Teilhard's good luck at Piltdown. Although records are frustratingly vague, I believe that all the Piltdown pieces were found by the original trio—Dawson, Smith Woodward, and Teilhard. (In the official version, a workman may have given Dawson the first piece in 1908.) Dawson, of course, unearthed most of the material himself. Smith Woodward, so far as I can tell, found only one cranial fragment. Teilhard, who spent less time at Piltdown than his two colleagues, was blessed. He found a fragment of the elephant tooth, a worked flint, and the famous canine.

People who have never collected in the field probably do not realize how difficult and chancy the operation is when fossils are sparse. There is no magic to it, just hard work. A tooth in a gravel pit is about as conspicuous as the proverbial needle in a haystack. The hoaxer worked hard on his Piltdown material. He filed the canine and painted it to simulate age. Apes' teeth are not easy to come by. If I had but one precious item, I would not stick it into a large gravel heap and then hope that some innocent companion would find it. It would probably be lost forever, not triumphantly recovered. I doubt that I would ever find it again myself, after someone else had mucked about extensively in the pile.

Teilhard described his discovery in the first letter to Oakley: "When I found the canine, it was so inconspicuous amidst the gravels which had been spread on the ground for sifting that it seems to me quite unlikely that the tooth could have been planted. I can even remember Sir Arthur congratulating me on the sharpness of my eyesight." Smith Woodward's recollection (from his last book of 1948) is more graphic:

> We were excavating a rather deep and hot trench in which Father Teilhard, in black clothing, was especially energetic; and, as we thought he seemed a little exhausted, we suggested that he should leave us to do the hard labor for a time while he had comparative rest in searching the rain-washed spread gravel. Very soon he exclaimed that he had picked up the missing canine tooth, but we were incredulous, and told him we had already seen several bits of ironstone, which looked like teeth, on the spot where he stood. He insisted, however, that he was not deceived, so we both left our digging to go and verify his discovery. There could be no doubt about it, and we all spent the rest of that day until dusk crawling over the gravel in the vain quest for more.

I also have some doubt about Teilhard's flint, for it is the only Piltdown item indubitably found in situ. All the other specimens either came from gravel heaps that had been dug up and spread upon the ground or cannot be surely traced. Now in situ can signify one of two things (and the records do not permit a distinction). It may mean that the gravel bed lay exposed in a ditch, cliff, or road cut—in which case, anyone might have stuck the flint in. But it may mean that Teilhard dug into the layer from undisturbed, overlying ground—in which case, he could only have planted the flint himself.

Again, I regard this argument only as suggestive, not as definitive. It is the weakest point of all, hence its place at the bottom of my list. Perhaps Teilhard was simply a particularly keen observer.

Conclusions

What shall we make of all this? I can only imagine three conclusions. First, perhaps Piltdown has simply deluded another gullible victim, this time myself. Maybe I have just encountered an incredible string of coincidences. Could all the slips in the letters have been innocent errors of an aging man; the *comme par exprès* merely a literary device; the failure to use his best argument a simple oversight; his conspicuous silence beyond a few fleeting and unavoidable mentions only an aspect of a complex personality that no one has fathomed; his profound embarrassment just another facet of the same personality; the elephant and hippo Dawson's property. . . ? I just can't believe it. Coincidences recede into improbability as more and more independent items coagulate to form a pattern. The mark of any good theory is that it makes coordinated sense of a string of observations otherwise independent and inexplicable. Let us then assume that Teilhard knew Piltdown was a hoax, at least from 1920.

We are left with two possibilities. Was Teilhard innocent in the field at Piltdown? Did he tumble to the hoax later (perhaps when he deciphered the inconsistencies in Piltdown 2)? Did he then maintain silence out of loyalty to Dawson who had befriended him or because he didn't wish to stir a hornet's nest when he was not completely sure? But why, then, did he try so hard to exonerate Dawson in the letters to Oakley? For Dawson had used him and played on his youthful innocence as cruelly as he had deceived Smith Woodward. And why did he write a series of slips and half-truths to Oakley that embody, as their only pattern, an attempt to extract himself alone? And why such intense embarrassment and such conspicuous silence if he had guessed right but had been too unsure to say so?

Alan Ternes, editor of *Natural History,* made the interesting suggestion that Teilhard, as a priest, might have heard of the hoax through a confession by Dawson that he could not subsequently reveal. I have not been able to ascertain whether Dawson was Catholic; I do not think that he was. But I am told that priests may regard statements of contrition by other baptized Christians as privileged information. This is the most sensible version I have heard of the hypothesis that Teilhard knew about the fraud but did not participate in it. It would explain his silence, his embarrassment, even the *"comme par exprès."* But, in this case, why would Teilhard have tried to construct such an elaborate and far-fetched theory of Dawson's innocence in his first letter to Oakley? Confession may have required silence, but surely not sheltering by falsehood. Any why the slips and half-truths for his own exoneration in the subsequent letters?

This leaves a third explanation—that Teilhard was an active coconspirator with Dawson at Piltdown. Only in this way can I make sense of the pattern in Teilhard's letters to Oakley, the 1920 article, the subsequent silence, the intense embarrassment.

This conclusion raises two final issues. First, to cycle back to my introduction, conspiracies have a tendency to spread. Once we admit Teilhard into the plot, should we not wonder about others as well? In fact, several knowledgeable people have strong suspicions about some young subordinates in the British Museum. I have confined my work to Teilhard's role; I think that others may have participated.

Second, what about motive? However overwhelming, the evidence cannot satisfy us without a reasonable explanation for why Teilhard might have done such a thing. Here I see no great problem, although we must recast Piltdown (at least from Teilhard's standpoint) as a joke that went too far, not as a malicious attempt to defraud.

Teilhard was not the dour ascetic or transported mystic that his publications sometimes suggest. He was a passionate man—a genuine hero in war, a true adventurer in the field, a man who loved life and people, who strove to experience the world in all its pleasures and pains. I assume that Piltdown was merely a delicious joke for him—at first. At Hastings, he was an amateur natural historian, with no expectation of a professional career in paleontology. He probably shared the attitude toward professionals so common among his colleagues—there but for the vagaries of life go I. Why do they have the fame, the reputation, and the cash? Why do they sit at their desks and reap rewards while we, with deeper knowledge born of raw experience, amuse ourselves? Why not play a joke to see how far a gullible professional could be taken? And what a wonderful joke for a Frenchman, for England at the time boasted no human fossils at all, while France, with Neanderthal and Cro-Magnon, stood proudly as the queen of anthropology. What an irresistible idea—to salt English soil with this preposterous combination of a human skull and an ape's jaw and see what the pros could make of it.

But the joke quickly went sour. Smith Woodward tumbled too fast and too far. Teilhard was posted to Paris to become, after all, a *professional* paleontologist. The war broke out, and Teilhard had to leave just as the last act to quell skepticism, Piltdown 2, approached the stage. Then Dawson died in 1916, the war dragged on to 1918, and professional English paleontology fell further and further into the quagmire of acceptance. What could Teilhard say by the war's end? Dawson could not corroborate his story. The jobs and careers of other conspirators may have been on the line. Any admission on Teilhard's part would surely have wrecked irrevocably the professional career he had desired so greatly, dared so little to hope for, and at whose threshold he now stood with so much promise. What could he say beyond *comme par exprès.*

Shall we then blame Teilhard or shall we forgive him? We cannot simply laugh and forget. Piltdown absorbed the professional attention of many fine scientists. It led millions of people astray for forty years. It cast a false light upon the basic processes of human evolution. Careers are too short and time too precious to view so much waste with equanimity.

But who among us would or could have come clean in Teilhard's position? Unfortunately, intent does not always correlate with effect in our complicated world—yet I believe that we must judge a man primarily by intent. If Teilhard had acted for malice or in hope of reward, I would have no sympathy. But I cannot view his participation as more than an intended joke that unexpectedly turned to a galling bitterness almost beyond belief. I think that Teilhard suffered for Piltdown throughout his life. I believe that he must have cried inwardly as he watched Smith Woodward and even Boule himself make fools of themselves—the very men who had befriended and taught him. Could the anguish of Piltdown have been on his mind, when he made the following pledge from the trenches during World War I?

> I have come, these days, to realize one very elementary fact: that the best way to win some sort of recognition for my ideas would be for me myself to attain, in the trust possible sense of the word, to a "sanctity" that will be manifest to others—not only because of the particular force God would then give to whatever good is in my aspirations and influence—but also because nothing can give me more authority over men than for them to see me as someone who speaks to them from close to God. With God's help, I must live my "vision" fully, logically, and without deviation. There is nothing more infectious than the example of a life governed by conviction and principle. And now I feel sufficiently drawn to and sufficiently equipped for, such a life.

Teilhard paid his debt and lived a full life; may we all do so well.

17 | A Reply to Critics

I CAN'T FEIGN either sad scholarly surprise or the wounded indignation of a friendly critic branded as a dishonest miscreant. I knew what would happen when I published the preceding essay. I have not been so certain of swift retribution since I hit that glorious game-winning triple one sunny afternoon in 1950 (on my stickball court, home runs cleared the opposite building, but triples went through the third story windows). If hell has no greater fury than a woman scorned, then true believers know no greater disillusion than a God humanized. Teilhard was an international cult figure during the late 1950s and 1960s. His star burns less brightly today, fickleness being the norm in matters of fashion, but a core of devotees still waves his banner and stands ready to crush underfoot any suggestion that Teilhard's behavior may have been less worthy (or more human) than the most rarefied notion of ethereal saintliness. Nothing I say will call off their dogs of war, but may I reiterate for others more disposed to listen: honest to God, I am not out to destroy Teilhard. I think that he was a complex and fascinating man—far more inspiring as a real human than as the piece of celestial cardboard touted by his devotees. Also, though it is obviously not for me to say, I really do forgive him if he did what I suspect. He was young; he did not act for profit, either monetary or personal; he suffered; he maintained steadfast and admirable loyalty to all involved; he made no excuses.

Having thus unburdened, I shall proceed, in formulating this reply, to ignore most of the personal and vituperative commentary directed at me. I shall also withhold comment upon the larger volume of supportive and friendly letters—except to say, "Thanks so much for understanding what I tried to do."

The serious negative commentary came in two waves. For six to nine months after my article appeared in August 1980, I received replies that provided no new information, but gave different interpretations (upholding Teilhard's innocence) to the same data I presented—usually with arguments that I had anticipated and (at least to my own satisfaction) countered in the original article. The three most interesting pieces in this mode—by Professors Dodson, Washburn, and von Koenigswald—were published along with my response in *Natural History* for June 1981. I have not reprinted them here, both because I do not wish to burden this volume with too much of my own private passion in this matter, and because I do not feel that either the comments or my response added anything substantive to the debate. But people with a special interest in the subject, particularly those who do not share my opinion, should consult this exchange and not take my word for it.

In the second wave, rebuttals based on new information finally began to appear. I shall discuss here every substantive point that has come to my attention. I believe that the intense scrutiny devoted to my case has so far failed to weaken it—though readers must judge for themselves whether this claim merely reflects my own blind egoism, or a judicious account of the situation.

OTHER INTERPRETATIONS OF TEILHARD'S LETTERS TO OAKLEY

As one of my two strong points, I argued that Teilhard made a crucial slip in dates when he claimed that Dawson had pointed to the site where he had already discovered the remains of Piltdown 2. But, in the "official" chronology, this second find occurred in 1915, while Teilhard was mustered into the French army in 1914 and never saw Dawson again. I discussed three "innocent" interpretations of this slip in my original article, and gave my reasons for preferring a fourth reading—that Teilhard knew about Piltdown 2 because he had either planned or discussed this future episode with Dawson before he left.

Mary Lukas, a biographer of Teilhard and my most persistent critic, offered a first substantive rebuttal in immediate reaction to my article. She charged that Kenneth Oakley and I and all who ever read the Oakley letters had uniformly misinterpreted them. She claimed that Teilhard was referring not to the second Piltdown site—the one that supposedly yielded fossils in 1915—but to a second pit at the first site (which could have been excavated by 1913). But this cannot be because each of three times that Teilhard mentions this second find, he refers to it explicitly as the place "where the two small fragments of skull and the isolated molar were supposedly found in the rubble." Only one place yielded two skull fragments and a molar: the second *site,* "discovered" by Dawson in 1915. Since Lukas has since written several pieces attacking my hypothesis, but has never raised this charge again, I assume that she now recognizes its invalidity.

Ms. Lukas's major article appeared in the Jesuit journal, *America,* for May 23, 1981. It is primarily a detailed—and quite correct—argument for a different substantive point. In short, she spends most of her article demonstrating that Dawson probably took Teilhard to the site of Piltdown 2 in 1913. She characterizes my case in the following words:

> Since in his letter to Oakley Teilhard seemed to demonstrate that he had prior knowledge of Dawson's plans by admitting he had seen the second Piltdown site before anyone else claimed to have seen it, Teilhard, Mr. Gould continued, must have been guilty with Dawson at Piltdown.

Having thus portrayed my case, and then demonstrating that Dawson did show the second site to Teilhard, Lukas concludes her rebuttal:

> Teilhard could well have seen Site 2, the plough field of Sheffield Park, just at the time he would later tell Oakley he thought he had: in the summer of 1913.

I trust that careful readers of the original essay will realize that Lukas completely misrepresents me, and that I have never doubted that Dawson showed Teilhard the site of Piltdown 2 in 1913. I said so explicitly in my original article: "*Teilhard did visit the second site with Dawson in 1913, but they did not find anything. Dawson 'discovered' the skull bones at Piltdown 2 in January 1915, and the tooth not until July 1915.*"

It is a clear matter of public record that Dawson showed the second site to several people before 1915, not only to Teilhard in 1913, but also to Smith Woodward several times in 1914 (as also mentioned in my article). But these trips led to *no fossil discoveries.* The bones of Piltdown 2 were "officially" unearthed in 1915, after Teilhard left England to join the French army. Yet Teilhard claimed knowledge of the finds before his departure. Lukas has spent a great deal of effort demonstrating something that everyone knows and admits—and that has no relevance to the case.

If these two claims worried me not at all, a third raised in April 1981 initially seemed far more serious. Indeed, I was quite willing to recant this part of my argument (and thus seriously compromise the entire case), if the claim could be substantiated as made. In short, Dr. J. S. Weiner, one of the original Piltdown debunkers and author of a fine book setting out the case for Dawson's complicity (*The Piltdown Forgery,* Oxford University Press, 1955), presented a lecture at Georgetown University as part of a celebration for the centenary of Teilhard's birth. He brought with him a previously unpublished letter from Dawson to Smith Woodward dated July 3, 1913. Reports of the lecture (I was not invited and did not attend) held that the letter spoke of fossil finds—not merely fruitless visits—at the second site in 1913. In fact, the 1913 letter supposedly reported the discovery of a skull fragment that later became part of the Piltdown 2 material. Now, if Dawson actually "found" material at Piltdown 2 in 1913, he might well have mentioned it to Teilhard (why not, since he had already written to Smith Woodward), and my case would evaporate. (I could not, after all, charge Teilhard for his claim that he had seen all three items of Piltdown 2, when Dawson had told him of the skull fragment alone. A clear memory of *some* fossil material from Piltdown 2 could easily be conflated, forty years later, with the entire later find.)

Thus, I approached the archives of the British Museum (where the original letter resides) in February 1982 with a strong sense of trepidation and humility. But I soon found that the July 3 letter does not speak about the material of Piltdown 2 at all; the smoking gun turned out to be a red herring. The letter reads:

> My dear Woodward
>
> I have picked up the frontal part of a human skull this evening on a plough field covered with flint gravel. It is a new place, a long way from Piltdown, and the gravel lies 50 feet below level of Piltdown, and about 40 to 50 feet above the present river base. It is *not* a *thick* skull but it may be a descendant of Eoanthropus. The brow ridge is slight at the edge, but full and prominent over the nose. It was coming on dark and raining when I left the place but I have marked the spot. . . . [Dawson's italics]

Now the material of Piltdown 2 does include two fragments of skull, one a frontal—and one is considerably thinner than the distinctive and remarkably thickened skull pieces of Piltdown 1. But the thin fragment of Piltdown 2 is an occipital, not a frontal (that is, a piece of the back, not the front, of the skull). (It is also cleverly cut to imitate the thinnest portion of the Piltdown 1 skull, and therefore not to arouse suspicion for its differences.) The frontal fragment of Piltdown 2 is as thick as the skull material from Piltdown 1—indeed (in restrospect) it *is* part of the same skull used to construct the original forgery. Thus, the *thin* frontal fragment described by Dawson in 1913 is not part of the Piltdown 2 material. In fact, Dawson himself recognized the differences in his 1913 letter and speculated that the thin frontal fragment might be a "descendant of *Eoanthropus*" (*Eoanthropus* was the official taxonomic designation of Piltdown man).

This resolution pleased me because the tale of the 1913 letter never made much sense. If Dawson had actually found a skull piece of Piltdown 2 in 1913, then why did he rereport it in his first explicit letter about Piltdown 2 on January 9, 1915? I recognize, of course, that this resolution leaves us with another interesting mystery—namely, what ever happened to the thin frontal fragment described in the 1913 letter? For it is never mentioned again so far as I know, and it forms no part of the Piltdown lode. I have no idea, and know no source of potential evidence. I would conjecture, however, that Dawson showed it to Smith Woodward, and that they judged it to be just what Dawson had suggested—a descendant of Piltdown, in fact a fragment from a modern skull—and therefore paid no further attention to it. Perhaps, for once, Smith Woodward didn't fall, and Dawson chose not to press his luck.

I do, of course, recognize that one could still construct, from the 1913 letter correctly interpreted, a scenario for Teilhard's exoneration: Dawson told Teilhard about the 1913 fragment; Teilhard knew at the time that it had nothing to do with a second site coeval with Piltdown 1; he also knew that it did not come from the site of Piltdown 2 that he had visited with Dawson in 1913; he forgot all this later, remembered that something else had been found somewhere in 1913, and confused this information with the later find of Piltdown 2. But this degree of special pleading makes a weak and conjectural case. It is surely no stronger than the simpler and related claim that I discussed and rejected in the original article—that Teilhard saw only the site of Piltdown 2, but misremembered forty years later and thought that Dawson had told him about the fossils as well.

THE NATURE OF TEILHARD'S LIFE AT ORE PLACE

Under the classical rubric of "opportunity," several critics familiar with Jesuit life have charged that Teilhard was so restricted by the rules of his seminary that he could not have spent sufficient private time with Dawson to hatch and execute such a plot.

Karl Schmitz-Moorman, editor of the excellent facsimile edition of Teilhard's technical works, raised two arguments on this theme (*Teilhard Newsletter,* vol. 14, July, 1981; Ms. Lukas reiterates them in a related article in the same issue, as did Thomas M. King, S.J. of Georgetown University in a private communication to me). First, Schmitz-Moorman argues, strict Jesuit rules kept Teilhard virtually confined to quarters or chaperoned by other Jesuits when outside—in other words, no opportunity for private plotting:

> Teilhard was always under supervision when he was working in the field during the seminary years. The same was true for his life inside the seminary. Doors could be opened at any time and superiors could step in to see how the students were getting on. Rules were very strict. [p. 3]

Second, Schmitz-Moorman reminds us that Teilhard was not a frequent visitor to the actual site of Piltdown 1: "When Teilhard left England in the summer of 1912 to begin his studies in Paris, he had been only once to Piltdown" (p. 3). He also excavated there with Dawson and Smith Woodward twice during August, 1913. Ore Place, Teilhard's seminary, was located some forty miles from Piltdown 1.

I must reject the premise of the second argument and the claim of the first. "In order to have taken part in the Piltdown hoax," Schmitz-Moorman argues, "Teilhard would have had to make many visits to Uckfield in East Sussex" (p.

3)—the site of Piltdown 1. But why? Must all conspirators pull the trigger itself? I have never charged Teilhard with putting the actual bits in the ground; I have always assumed that Dawson played this role. There were so many other things to do—getting, breaking, and doctoring specimens just for starters.

Schmitz-Moorman's first argument reminds me of Casey Stengel's immortal distinction between general categories and specific cases. (When asked why he blew the Mets's first draft choice on a particularly inept catcher, Stengel remarked: "If you don't have a catcher, you're likely to have a lot of passed balls.") Same problem (in reverse) with Schmitz-Moorman. I do not doubt his generality about life in Jesuit seminaries. But the specific record for Ore Place indicates that Teilhard had more than enough freedom to work with Dawson. First of all, his own letters (see bibliography) speak of a frequency and range of excursions far in excess of what the general rules would allow. Second, the standard biography of Teilhard (*Teilhard de Chardin* by Claude Cuénot, p. 12) states:

> Thanks to the liberal attitude of the Rector at Hastings, Teilhard was allowed to go more frequently on scientific walks and excursions, finding specimens to offer to the British Museum or the Museum at Hastings. He had now advanced beyond the amateur class, and was manifesting a clear bent towards the paleontology of the vertebrates.

NATURE OF THE RELATIONSHIP BETWEEN DAWSON AND TEILHARD

Peter Costello, an independent researcher in Dublin and author of a forthcoming book about Piltdown, found previously unpublished letters from Teilhard to Dawson in the archives of the British Museum (Natural History). Costello (1981, see bibliography) published one of the letters, suggesting that its tone provided a refutation of my hypothesis.

I quote the entire letter and Costello's interpretation. Teilhard wrote to Dawson on July 10, 1912:

> Dear Mr. Dawson,
>
> I am sorry to tell you that it is impossible for me to go to Lewes, next week, because I have to start from Hastings on Tuesday! I hope, nevertheless, that we will again dig together the Uckfield's gravel: next year, I am likely to study Natural History in France, and to spend my holidays in England. If so, I will surely do my best to see you. Until I give you my definitive address, you can write to me at: 'Château de Sarcenat, par Oreines, Puy-de-Dôme.'
>
> I am very thankful to you for your kindness towards me during this last four years. Lewes will certainly be one of my best remembrances of England, and you may be sure that I shall often pray God to bless the Castle Lodge [Dawson's residence].
>
> Yours sincerely,
> P. Teilhard

Costello concludes (1981, pp. 58–59):

> I suggest that this farewell letter, so touching in its expression of thanks, demonstrates (as do the others in the series), that the relationship between Dawson and Teilhard was one of mentor and pupil and that no conspiracy existed between them.

I do not see how this letter speaks either for or against my case. If one has been seeing too many old-style gangster movies and develops a peculiarly cardboard view of conspiracies, then I suppose that each participant might have to be slimy, unregenerate, utterly unkind, and devoid of any admirable quality. But I rather suspect that conspirators are not so far from a cross section among all of us. I do not see why they should not show loyalty to each other, pay thanks for kindnesses rendered, and even show deference to large differences in age and experience. Must conspirators be equals? Have no "mentor and pupil" ever plotted together? Shall we exonerate Teilhard because he and Dawson weren't on a first-name basis in formal, just post-Edwardian England? The letter is touching and it does reflect Teilhard in an admirable light. People are complex, with faults and virtues aplenty. I have always argued that Teilhard's virtues outweighed the major fault I have tried to identify.

Costello implies (in the statement quoted above) that his evidence is multiple and that "others in the series" confirm the tone of his quoted letter. Perhaps I did not search assiduously enough in the British Museum archives, but I could find only one other letter from Teilhard to Dawson, a short and uninformative piece of June 21, 1912, simply telling Dawson of his imminent departure and urging a visit to select fossils from Teilhard's collection for the British Museum. In other words, I think that Costello has quoted everything he has.

But I also made some discoveries of my own about the relationship between Dawson and Teilhard as reflected in letters of the British Museum archives. If the archives contain a paucity of Teilhard's letters, they are rich in Dawson's —and these letters belie Costello's chief claim that Dawson and Teilhard had only a passing and formal acquaintance. The high density of references to Teilhard in Dawson's letters (mostly to Smith Woodward) points both to a strong relationship between the two men and, especially, to a particular solicitude on Dawson's part towards Teilhard.

Consider, for example, a series of letters from Dawson to Smith Woodward in 1915, after Teilhard had left for the front. On March 9, Dawson writes: "I enclose a P.C. from Teilhard at the French front. He, no doubt, would be glad of any little bits of literature which you can send him." On April 3, he states: "Teilhard has now been moved to near the back of the English line in Flanders. He says he is all right 'body and mind'." And on July 3, in the same letter that reported the "discovery" of the Piltdown 2 molar, Dawson announced: "Teilhard wrote yesterday—he is quite well and in a quiet spot at present." (Unfortunately, no one has uncovered any of this correspondence between Dawson and Teilhard. I, for one, would dearly love to know what it contained). I submit that this degree of contact indicates a level of friendship and mutual concern far greater than that allowed by my critics. Limited contact is as crucial to their case as this demonstrated bond is to mine.

The prewar correspondence of Dawson and Smith Woodward shows a similar pattern. Six letters between October 1909 and October 1911 mention Teilhard and his work in collecting fossils. The pace picks up following Dawson's first notice of the Piltdown skull to Smith Woodward on February 14, 1912. Six more letters mention Teilhard between then and November 21, 1912.

I believe, on this point and others, that all my critics have used a peculiar style of argument amounting to an a priori refusal to consider my case seriously—that is, Costello, Lukas, and Schmitz-Moorman all argue that I must be wrong because the written record provides no direct evidence for conspiracy. Costello wrote to me (September 4, 1981): "Nowhere can one read anything that suggests they were plotting together." Lukas writes (1981 in bibliography, p. 426): "According to his letters, both published and unpublished, to family and friends, Teilhard's relationship to Dawson was anything but close." And further on: "Before the Piltdown adventure began Teilhard and Dawson seem to have met only four times."

But surely, if any principle regulates conspiracy, we may state that plotters do not generally write extensive, contemporary accounts of their deeds (later confessions for profit or expiation notwithstanding). If Teilhard and Dawson were plotting, their machinations would certainly not have appeared in letters to parents and friends, or in preserved letters to each other. To identify conspiracy, one must search for implicit pattern behind the stated record, not for explicit contemporary confessions.

Thus, in conclusion, I believe that no strong arguments have been raised against my case and that, in one area, I have bolstered my account by recognizing the extent of

Dawson's concern for Teilhard as expressed in his letters to Smith Woodward. Moreover, for all the criticism of my first strong point (the letters to Oakley), my detractors have been conspicuously silent about my second strong argument (Teilhard's pattern of silence concerning Piltdown in his extensive publications on human evolution). The more I think about this, the more it becomes, in Alice's immortal words, "curiouser and curiouser."

Since writing the original article, another small point, making Teilhard's silence even more puzzling, has come to my attention. When Peking man was discovered, its cranium was reconstructed incorrectly to yield a capacity lying, like Piltdown's, in the modern human range. This unleashed a flurry of commentary about the relationship between Piltdown and Peking. Now Teilhard was in China where he was contributing (as a geologist) to the original Peking finds. He was the only one there with personal knowledge of Piltdown. Yet, so far as I can tell, he said nothing at all. His own mentor, Marcellin Boule, published a paper comparing the Peking and Piltdown crania. It included long quotations from Teilhard about the geology of the Peking site, but not a word from him about the crania.

Again I repeat, if Teilhard considered the Piltdown material to be genuine, the skull[1] provided his strongest direct evidence for the postulate that he held most dear and that motivated all his work on man's spiritual evolution—multiple parallel lineages ascending toward the domination of spirit over matter. And he never mentioned Piltdown beyond a half-dozen quick, unavoidable, and almost embarrassed references. Why?

Lest readers think that all speculation on Piltdown has gone against Teilhard's involvement of late, I add that Dr. L. Harrison Matthews, one of the grand old men of British zoology (see essay 11) and a personal acquaintance of nearly

---

[1]Even if he decided that the jaw belonged to an ape and had been mixed by accident into the two Piltdown sites, the skull remained a genuine human fossil—thick and therefore "primitive," but of modern human capacity. This is the dual solution that he favored in his 1920 article.

everyone involved in the original case, has published his novel-length reconstruction in the *New Scientist* (see bibliography). He sees the necessity of Teilhard's involvement, but develops a highly complicated scenario in which Dawson begins the hoax alone, Teilhard then recognizes what Dawson is doing and, to let Dawson know and warn him off any future hoaxing, Teilhard manufactures, plants, and finds the canine himself. The war then intervenes, Dawson dies, and Teilhard is backed into a corner of silence. I welcome this basic insight that Teilhard cannot be excluded, but regard his case as too complex in that most difficult of ways—to be right, each of two dozen unsubstantiated events must break exactly in Harrison Matthews's way. I continue to urge the simpler view—that Teilhard worked with Dawson at least from 1912 until he left for the front.

In his last public comment on Piltdown, Kenneth Oakley, who died on November 2, 1981, wrote a letter to the *New Scientist* (published posthumously on November 12, 1981) stating his disagreement with Harrison Matthews. I do not know what opinion he held of my case at the time of his death. After my original article, he wrote, by invitation, a letter to *The Times* (London) stating that, in the absence of definite proof, Teilhard should be given the benefit of the doubt. (Were I a judge, and this a legal proceeding with standards so necessarily different from historical inquiry, I would have to concur. Of my original article, one close friend remarked that I had established the grounds for an indictment, but not for a conviction.) I have also seen statements from a private letter in which Oakley, arguing from the 1913 letter of Dawson to Smith Woodward, rejects my first claim based on the letters between Teilhard and Oakley, but explicitly does not state a belief in Teilhard's innocence. (I have already indicated why I think the 1913 letter is irrelevant to my case.) It is a matter of record among several of Oakley's close friends that he long maintained private suspicions of Teilhard's active involvement in at least some aspect of the case.

I bring this up because some critics have charged me with dishonesty in imputing a more favorable view to Oakley than he actually held. I can only state that I sent a copy of the original article to Oakley before it was published, asking directly if I had represented him accurately and if he approved my attribution. He wrote to me (on June 6, 1980): "I read straight through your paper without finding anything (of any importance) which I would wish you to alter."

As a final comment, I must express mixed feelings two years after the original article. I am delighted to find my hypothesis strong and undiminished (admittedly in my biassed view) by a series of searching and intensely negative commentaries. On the other hand, I confess that I held secret hopes, nurtured perhaps by my own overly heroic view of life. I hoped that some old man would come down off a mountain or out of a monastery bearing a yellowed document of confession from Teilhard. Or that some trusted friend would open a bank vault and make public the "letter to be read either at the 100th anniversary of my death or when someone figures out my involvement in Piltdown." Nothing like this has happened. No good arguments have been raised against me, but I must admit that nothing of great consequence has turned up in my favor either. I began the first essay that I wrote on Piltdown (reprinted in *The Panda's Thumb*) before I was much interested in Teilhard's role, with the words: "Nothing is quite so fascinating as a well-aged mystery." And so Piltdown remains, though I might add that nothing would be quite so satisfying as a definitive resolution.

18 | Our Natural Place

WHEN LINNAEUS SOUGHT to classify all of life in 1758, he called his great work the *Systema Naturae,* the "System of Nature." Biologists of all subsequent generations have flooded the scientific literature with alternative, but equally comprehensive, systems. The content changes, but the passion for building systems remains. Our urge to make sense of the complexity that surrounds us, to put it *all* together, overwhelms our natural caution before such a daunting task.

A curious irony infects this tradition of building comprehensive systems in biology. Biologists present their systems either as necessary truths of superior logic or as ineluctable conclusions drawn from unrivaled powers of observation—in other words, as objective renderings of nature, heretofore unappreciated. In fact, these systems share only one common property—and it is neither objectivity nor superior wisdom. They are, at base, attempts to resolve a (perhaps *the*) cardinal question of intellectual history: What is the role and status of our own species, *Homo sapiens,* in nature and the cosmos?

Systems follow one of two strategies in their attempt to make sense of "man's place in nature," to use T. H. Huxley's phrase. One strategy, the "picket fence" in my terminology (see essay 12 in *The Panda's Thumb*), devises a pervasive order for the rest of nature, but separates humans alone with a brand of superiority. Thus, Charles Lyell envisaged a world ever churning and changing, but always remaining the same—replacement without improvement. Only man, a recent imposition of moral perfection upon a stable world, broke the pattern of change without progress. A. R. Wallace attributed all features of organisms to the molding power of natural selection—except for one product of divine inspiration: the human brain.

The second strategy takes an opposite tack in pursuit of the same goal—a placement within nature that will make some sense of our lives. This strategy argues for no separation between man and nature at all. These theories of continuity can proceed in either direction, and I shall discuss one recent example of each as representatives of a long tradition of flawed argument. The first view—I shall call it zoocentric—builds general principles from the behavior of other animals and then subsumes humans completely into the rubric because we are, after all and undeniably, animals too. The second view—I shall call it anthropocentric—tries to subsume nature in us by viewing our peculiarities as the goal of life from the start.

Evolutionary theory itself has an appropriately zoocentric core. General principles cannot arise from the behavior of a single species, yet all species must conform to the principles. The role of this mild zoocentricity in breaking down the sturdy picket fences that existed before Darwin's time ranks as a great event in the history of human thought. But the zoocentric view can be extended too far into a caricature often called the "nothing but" fallacy (humans are "nothing but" animals).

The simplistic accounts of human sociobiology now flooding popular literature embody this overextended version of zoocentrism. Sociobiology is not just any statement that biology, genetics, and evolutionary theory have something to do with human behavior. Sociobiology is a specific theory about the nature of genetic and evolutionary input into human behavior. It rests upon the view that natural selection is a virtually omnipotent architect, constructing organisms part by part as best solutions to problems of life in local environments. It fragments organisms into "traits," explains their existence as a set of best solutions, and argues that each trait is a product of natural selection operating "for" the form or behavior in question. Applied to humans, it must view *specific* behaviors (not just general potentials) as adaptations built by natural selection and rooted in genetic determinants, for natural selection is a theory of genetic change. Thus, we are presented with unproved and unprovable speculations about the adaptive and genetic basis of specific human behaviors: why some (or all) people are aggressive, xenophobic, religious, acquisitive, or homosexual.

Zoocentrism is the primary fallacy of human sociobiology, for this view of human behavior rests on the argument that if the actions of "lower" animals with simple nervous systems arise as genetic products of natural selection, then human behavior should have a similar basis. Humans are animals too, aren't we? Yes, but animals with a difference. And that difference arises, in part, as a result of enormous flexibility based on the complexity of an oversized brain and the potentially cultural and nongenetic basis of adaptive behaviors—aspects of human construction that debar any zoocentric extrapolation from why some insects eat their mates to murder in human families.

Ironically, the zoocentrism of human sociobiology is often an illusion hiding a precisely opposite mode of reasoning. I argued previously that "objective" systems are often unconscious masquerades that reflect our own prejudices and hopes imposed upon nature. Much of human sociobiology trades upon the idea that if distinctive human behaviors can be found, albeit in rudimentary form, among "lower" animals, then these behaviors must be "natural" in humans too, a product of biological evolution. Sociobiologists are often fooled by misleading external and superficial similarity between behaviors in humans and other animals. They attach human names to what other creatures do and speak of slavery in ants, rape in mallard ducks, and adultery in mountain bluebirds. Since these "traits" now exist among "lower" animals, they can be derived for humans as natural, genetic, and adaptive. But they never did exist outside a human context. If mallard males seem to force physically weaker females to copulate, what possible relationship, beyond meaningless superficial appearance, can such an act have with human rape? No one can argue that the two behaviors are truly homologous—that is, based on the same genes inherited from a common ancestor. If the similarity is meaningful, it can only be analogous—that is, reflecting different evolutionary origins but performing the same biological function. Yet mallard behavior is part of the normal repertoire and seems to have evident utility in increasing the reproductive fitness of males; while human rape is a social pathology rooted in power and powerlessness, not in sex and reproduction.

Is this not mere pedantic grousing? Aren't the human terms a cute, graphic, and acceptable shorthand for what we all recognize as a more complex reality? Not when a colleague describes aggressive responses of male mountain bluebirds toward other males in the vicinity of its nest with these words: "The term 'adultery' is unblushingly employed . . . without quotation marks, as I believe it reflects a true analogy to the human concept. . . . It may also be prophesied that continued application of a similar evolutionary approach will eventually shed considerable light on various human foibles as well."

Such an old story. We hold a mirror to nature and see ourselves and our own prejudices in the glass. Historical examples abound. Aristotle described the large bee that leads the swarm as a "king," and this misidentification of the only sexual female around persisted for nearly two thousand years, at least to an Elizabethan madrigal I sang last week: "I do love thee as the spring/Or the bees their careful king" (careful, that is, in the old sense of caring, not the modern meaning of cautious). Zoocentric systems fail primarily because they never are what they claim to be. The "objective" animal behavior, under which they subsume human acts, is an imposition of human preferences from the start.

The more venerable, anthropocentric systems at least have the virtue of explicit self-recognition. They take Protagoras seriously in his claim that "man is the measure of all things," and falter only in the hubris of arguing that evolution undertook its elaborate labor of some 3½ billion years only to generate the little twig that we call *Homo sapiens.* Anthropocentric systems have been out of vogue among scientists, at least in England and America, since Darwin's time, but one version enjoyed some spectacular popularity a few years back—the system of a man discussed in quite a different context elsewhere in this section, the Jesuit priest and distinguished paleontologist Pierre Teilhard de Chardin.

When Teilhard died in 1955, his evolutionary speculations, long suppressed by ecclesiastical authority, saw light, and his best seller, *The Phenomenon of Man,* became a cult book of the 1960s. Teilhard's florid and mystical writing is often more difficult to decipher than his role at Piltdown, but I believe that the general line of his argument can be simply expressed. (A convinced Teilhardian might brand me as a shallow, heartless scientist, unable to appreciate the profundity of Teilhard's vision. But difficult, convoluted writing may simply be fuzzy, not deep. Teilhard's vision is rich in scope and tradition—for it is an old argument clothed in new terminology—but the essence of his position can still be stated with everyday words.)

Teilhard believed that evolution proceeds in a definite and irreversible direction. To understand the nature of that movement, we do not look back to the origin of life and its physical properties, but to the latest product—to *Homo sapiens* itself. For life has been moving in our direction from the start. The advance of life records an ever increasing domination of spirit over matter. This ineluctable increase in consciousness can be grasped by studying two of its material products: among lower animals, diffuse and simple nervous systems evolve into centralized organs (brains) with subsidiary parts; among higher animals, brains increase in size and complexity throughout evolution. In an autobiographical essay, Teilhard wrote:

> I never really paused for a moment to question the idea that the progressive spiritualization of matter—so clearly demonstrated to me by paleontology—could be anything other, or anything less, than an irreversible process. By its gravitational nature, the universe, I saw, was falling—falling forwards—in the direction of spirit as upon its stable form. In other words, matter was not ultra-materialized as I would at first have believed, but was instead metamorphosed in psyche.

Human evolution is the culmination of this psychic advance. In the anthropocentric vision, life only makes sense in terms of its striving toward man. We are inextricably part of nature because nature has been yearning toward us from the start. In a 1952 manuscript on human socialization, Teilhard stated:

> Human evolution is nothing else but the natural continuation, at a collective level, of the perennial and cumulative process of "psychogenetic" arrangement of matter which we call life. . . . The whole history of mankind has been nothing else (and henceforth it will never be anything else) but an explosive outburst of ever-growing cerebration. . . . Life, if fully understood, is not a freak in the universe—nor man a freak in life. On the contrary, life physically culminates in man, just as energy physically culminates in life.

Since evolution follows a directed path, the tree of life is not a randomly ramifying network, but a bundle of branches, tied by genealogy at their base, diverging during their history, yet always moving in the same basic direction. The energy of matter compels divergence; the force of increasing consciousness imposes a common upward advance. Related species should form a set of *multiple, parallel lineages,* each diverging and adapting to a local environment, but each gaining continually in its spirit/matter ratio. Teilhard wrote in 1922 that "evolution . . . resolves itself into innumerable lines which diverge at such length that they appear parallel."

With the appearance of man, evolution has reached a crux. Spirit has accumulated far enough to reach self-consciousness. Indeed, a new layer has appeared in the earth's concentric structure. Teilhard praised the great Austrian geologist Eduard Suess for introducing the term "biosphere" as an addition to the traditional concentric layers of lithosphere and atmosphere. But consciousness, Teilhard argued, has added yet another layer: "the psychically reflexive human surface . . . the noosphere."

Teilhard describes the noosphere as a physical reality, as a thin and fragile layer, now spread throughout the earth following the emergence of human ancestors from Africa and their subsequent migration to all continents. He wrote in a 1952 manuscript: "Above the old Biosphere there is spread now a 'Noosphere'. As to the material reality of this enormous event, there is nobody who will disagree." In a posthumous essay published four years later, he described the noosphere as "the marvelous sheet of humanized and socialized matter, which, despite its incredible thinness, has to be regarded positively as the most sharply individualized and the most specifically distinct of all the planetary units so far recognized."

The emergence of a noosphere, now so thin and fragile, represents the turning point of universal evolution. Teilhard wrote in 1930:

> The phenomenon of Man represents nothing less than a general transformation of the earth, by the establishment at its surface of a new layer, the thinking layer, more vibrant and more conductive, in a sense, than all metal; more mobile than all fluid; more expansive than all vapor. . . . And what gives this metamorphosis its full grandeur is that it was not produced as a secondary event or a fortuitous accident—but in the form of a

turning point essentially foreordained, from the beginning, by the nature of the general evolution of our planet.

Evolution has now reached its halfway point. Heretofore, despite the progressive gain of spirit, matter has dominated and evolutionary lineages, although moving in the same general direction, have constantly diverged. But the noosphere marks the beginning of the dominion of spirit over matter. As spirit gains the upper hand, convergence shall begin. The fragile noosphere shall thicken. The direction of a billion years shall be reversed, and conscious lineages (within *Homo sapiens* at least) will begin to converge as spirit gains rapidly in its sway over matter.

Convergence has already begun in the process of human socialization. In terms of vulgar mechanics, human cultural evolution may be a process different from Darwinian biological evolution, but both participate in a higher unity as sequential aspects of universal direction. Human socialization, Teilhard writes, has engendered "a vast and specific process of physico-psychical convergency . . . whose sudden appearance and acceleration in the course of the last century is perhaps the most revolutionary event registered so far in human history. . . . The human world is decidedly caught, today and forever, in an irresistibly tightening vortex of unification."

The rush to convergence must concentrate and accelerate until all spirit, freer and freer of encumbering matter, amalgamates at a single point that Teilhard called Omega, identified with God, and, so far as I can tell, conceived as a reality, not a metaphor or symbol. He describes this apotheosis, graphically if not with perfect clarity, in *The Phenomenon of Man:*

> Convergent movement will attain such intensity and such quality that mankind, taken as a whole, will be obliged . . . to reflect upon itself at a single point . . . to abandon its organo-planetary foothold so as to pivot itself on the transcendent center of its increasing

> concentration. This will be the end and the fulfillment of the spirit of the earth.
>
> The end of the world: the wholesale internal introversion upon itself of the noosphere, which has simultaneously reached the uttermost limit of its complexity and centrality.
>
> The end of the world: the overthrow of equilibrium, detaching the mind, fulfilled at last, from its material matrix, so that it will henceforth rest with all its weight on God-Omega.

And so, evolution labored for billions of years, produced perhaps a hundred million species of plants, bugs, and worms along the way, all to achieve, through the agency of one species endowed with consciousness, the union of spirit with God in splendid concentration at the point Omega. All previous life existed for us and for what *we* could become. Like the floating fetus that embodies the promise of futurity at the close of *2001: A Space Odyssey,* we (or rather our thickening spirit layer, soaring upward) are the heirs and purpose of all previous life. This is the anthropocentric vision with a vengeance.

What can one say to such a scheme? Would it be too literal and mean spirited to argue that it seems to fail at its only points of testable contact with the fossil record? Few paleontologists can discern any general, much less inevitable, trend to increasing braininess in the history of life. Most animal species are insects, mites, copepods, nematodes, mollusks, and their cousins, and I, at least, can see no pervasive trend among them toward the domination of matter by spirit. And the evolutionary tree does look more to me like a complexly wandering and ramifying bush than a bundle of parallel twigs growing upward in a definite direction. Of course, I realize that Teilhard used the term evolution in a metaphysical sense to identify the laws of cosmic progress, not in our usual sense to specify the mechanics of organic change (which Teilhard recognized and studied, but called *transformisme*). Teilhard's technical works in paleontology are sound and solid, but they deal with *transformisme* and exist in a world of discourse quite separate from his anthropocentric vision of cosmic evolution.

Perhaps the problem with all these visions—zoocentric as well as anthropocentric—is our penchant for building comprehensive and all-encompassing systems in the first place. Maybe they just don't work. Maybe they must be defeated by the inherent complexity and ambiguity of our place in nature. How can we erect a picket fence when humans are so inextricably bound in nature? But how can we opt for complete continuity, either by working up from other animals (zoocentric) or down from humans (anthropocentric) when humans are so special, for better or for worse? We are but a tiny twig on a tree that includes at least a million species of animals, but our one great evolutionary invention, consciousness—a natural product of evolution integrated with a bodily frame of no special merit—has transformed the surface of our planet. Gaze upon the land from an airplane window. Has any other species ever left so many visible signs of its relentless presence?

We live in an essential and unresolvable tension between our unity with nature and our dangerous uniqueness. Systems that attempt to place and make sense of us by focusing exclusively either on the uniqueness or the unity are doomed to failure. But we must not stop asking and questing because the answers are complex and ambiguous. We can do no better than to follow Linnaeus's advice, embodied in his description of *Homo sapiens* within his system. He described other species by the numbers of their fingers and toes, their size and their color. For us, in place of anatomy, he simply wrote the Socratic injunction: Know thyself.

5 | Science and Politics

19 Evolution as Fact and Theory*

KIRTLEY MATHER, who died last year at age ninety, was a pillar of both science and Christian religion in America and one of my dearest friends. The difference of a half-century in our ages evaporated before our common interests. The most curious thing we shared was a battle we each fought at the same age. For Kirtley had gone to Tennessee with Clarence Darrow to testify for evolution at the Scopes trial of 1925. When I think that we are enmeshed again in the same struggle for one of the best documented, most compelling and exciting concepts in all of science, I don't know whether to laugh or cry.

According to idealized principles of scientific discourse, the arousal of dormant issues should reflect fresh data that give renewed life to abandoned notions. Those outside the current debate may therefore be excused for suspecting that creationists have come up with something new, or that evolutionists have generated some serious internal trouble. But nothing has changed; the creationists have presented not a single new fact or argument. Darrow and Bryan were at least more entertaining than we lesser antagonists today. The rise of creationism is politics, pure and simple; it represents one issue (and by no means the major concern) of the resurgent evangelical right. Arguments that seemed kooky just a decade ago have reentered the mainstream.

---

*First appeared in *Discover Magazine,* May 1981.

The basic attack of modern creationists falls apart on two general counts before we even reach the supposed factual details of their assault against evolution. First, they play upon a vernacular misunderstanding of the word "theory" to convey the false impression that we evolutionists are covering up the rotten core of our edifice. Second, they misuse a popular philosophy of science to argue that they are behaving scientifically in attacking evolution. Yet the same philosophy demonstrates that their own belief is not science, and that "scientific creationism" is a meaningless and self-contradictory phrase, an example of what Orwell called "newspeak."

In the American vernacular, "theory" often means "imperfect fact"—part of a hierarchy of confidence running downhill from fact to theory to hypothesis to guess. Thus, creationists can (and do) argue: evolution is "only" a theory, and intense debate now rages about many aspects of the theory. If evolution is less than a fact, and scientists can't even make up their minds about the theory, then what confidence can we have in it? Indeed, President Reagan echoed this argument before an evangelical group in Dallas when he said (in what I devoutly hope was campaign rhetoric): "Well, it is a theory. It is a scientific theory only, and it has in recent years been challenged in the world of science—that is, not believed in the scientific community to be as infallible as it once was."

Well, evolution *is* a theory. It is also a fact. And facts and theories are different things, not rungs in a hierarchy of increasing certainty. Facts are the world's data. Theories are structures of ideas that explain and interpret facts. Facts do not go away while scientists debate rival theories for explaining them. Einstein's theory of gravitation replaced Newton's, but apples did not suspend themselves in mid-air pending the outcome. And human beings evolved from apelike ancestors whether they did so by Darwin's proposed mechanism or by some other, yet to be discovered.

Moreover, "fact" does not mean "absolute certainty." The final proofs of logic and mathematics flow deductively from stated premises and achieve certainty only because they are *not* about the empirical world. Evolutionists make no claim for perpetual truth, though creationists often do (and then attack us for a style of argument that they themselves favor). In science, "fact" can only mean "confirmed to such a degree that it would be perverse to withhold provisional assent." I suppose that apples might start to rise tomorrow, but the possibility does not merit equal time in physics classrooms.

Evolutionists have been clear about this distinction between fact and theory from the very beginning, if only because we have always acknowledged how far we are from completely understanding the mechanisms (theory) by which evolution (fact) occurred. Darwin continually emphasized the difference between his two great and separate accomplishments: establishing the fact of evolution, and proposing a theory—natural selection—to explain the mechanism of evolution. He wrote in *The Descent of Man:* "I had two distinct objects in view; firstly, to show that species had not been separately created, and secondly, that natural selection had been the chief agent of change . . . Hence if I have erred in . . . having exaggerated its [natural selection's] power . . . I have at least, as I hope, done good service in aiding to overthrow the dogma of separate creations."

Thus Darwin acknowledged the provisional nature of natural selection while affirming the fact of evolution. The fruitful theoretical debate that Darwin initiated has never ceased. From the 1940s through the 1960s, Darwin's own theory of natural selection did achieve a temporary hegemony that it never enjoyed in his lifetime. But renewed debate characterizes our decade, and, while no biologist questions the importance of natural selection, many now doubt its ubiquity. In particular, many evolutionists argue that substantial amounts of genetic change may not be subject to natural selection and may spread through populations at random. Others are challenging Darwin's linking of natural selection with gradual, imperceptible change through all intermediary degrees; they are arguing that most evolutionary events may occur far more rapidly than Darwin envisioned.

Scientists regard debates on fundamental issues of theory as a sign of intellectual health and a source of excitement. Science is—and how else can I say it?—most fun when it plays with interesting ideas, examines their implications, and recognizes that old information may be explained in surprisingly new ways. Evolutionary theory is now enjoying this uncommon vigor. Yet amidst all this turmoil no biologist has been led to doubt the fact that evolution occurred; we are debating *how* it happened. We are all trying to explain the same thing: the tree of evolutionary descent linking all organisms by ties of genealogy. Creationists pervert and caricature this debate by conveniently neglecting the common conviction that underlies it, and by falsely suggesting that we now doubt the very phenomenon we are struggling to understand.

Secondly, creationists claim that "the dogma of separate creations," as Darwin characterized it a century ago, is a scientific theory meriting equal time with evolution in high school biology curricula. But a popular viewpoint among philosophers of science belies this creationist argument. Philosopher Karl Popper has argued for decades that the primary criterion of science is the falsifiability of its theories. We can never prove absolutely, but we can falsify. A set of ideas that cannot, in principle, be falsified is not science.

The entire creationist program includes little more than a rhetorical attempt to falsify evolution by presenting supposed contradictions among its supporters. Their brand of creationism, they claim, is "scientific" because it follows the Popperian model in trying to demolish evolution. Yet Popper's argument must apply in both directions. One does not become a scientist by the simple act of trying to falsify a rival and truly scientific system; one has to present an alternative system that also meets Popper's criterion—it too must be falsifiable in principle.

"Scientific creationism" is a self-contradictory, nonsense phrase precisely because it cannot be falsified. I can envision observations and experiments that would disprove any evolutionary theory I know, but I cannot imagine what potential data could lead creationists to abandon their beliefs. Unbeatable systems are dogma, not science. Lest I seem harsh or rhetorical, I quote creationism's leading intellectual, Duane Gish, Ph.D., from his recent (1978) book, *Evolution? The Fossils Say No!* "By creation we mean the bringing into being by a supernatural Creator of the basic kinds of plants and animals by the process of sudden, or fiat, creation. We do not know how the Creator created, what processes He used, *for He used processes which are not now operating anywhere in the natural universe* [Gish's italics]. This is why we refer to creation as special creation. We cannot discover by scientific investigations anything about the creative processes used by the Creator." Pray tell, Dr. Gish, in the light of your last sentence, what then is "scientific" creationism?

Our confidence that evolution occurred centers upon three general arguments. First, we have abundant, direct, observational evidence of evolution in action, from both field and laboratory. This evidence ranges from countless experiments on change in nearly everything about fruit flies subjected to artificial selection in the laboratory to the famous populations of British moths that became black when industrial soot darkened the trees upon which the moths rest. (Moths gain protection from sharp-sighted bird predators by blending into the background.) Creationists do not deny these observations; how could they? Creationists have tightened their act. They now argue that God only created "basic kinds," and allowed for limited evolutionary meandering within them. Thus toy poodles and Great Danes come from the dog kind and moths can change color, but nature cannot convert a dog to a cat or a monkey to a man.

The second and third arguments for evolution—the case for major changes—do not involve direct observation of evolution in action. They rest upon inference, but are no less secure for that reason. Major evolutionary change requires too much time for direct observation on the scale of recorded human history. All historical sciences rest upon inference, and evolution is no different from geology, cosmology, or human history in this respect. In principle, we cannot observe processes that operated in the past. We must infer them from results that still surround us: living and fossil organisms for evolution, documents and artifacts for human history, strata and topography for geology.

The second argument—that the imperfection of nature reveals evolution—strikes many people as ironic, for they feel that evolution should be most elegantly displayed in the nearly perfect adaptation expressed by some organisms—the camber of a gull's wing, or butterflies that cannot be seen in ground litter because they mimic leaves so precisely. But perfection could be imposed by a wise creator or evolved by natural selection. Perfection covers the tracks of past history. And past history—the evidence of descent—is the mark of evolution.

Evolution lies exposed in the *imperfections* that record a history of descent. Why should a rat run, a bat fly, a porpoise swim, and I type this essay with structures built of the same bones unless we all inherited them from a common ancestor? An engineer, starting from scratch, could design better limbs in each case. Why should all the large native mammals of Australia be marsupials, unless they descended from a common ancestor isolated on this island continent? Marsupials are not "better," or ideally suited for Australia; many have been wiped out by placental mammals imported by man from other continents. This principle of imperfection extends to all historical sciences. When we recognize the etymology of September, October, November, and December (seventh, eighth, ninth, and tenth), we know that the year once started in March, or that two additional months must have been added to an original calendar of ten months.

The third argument is more direct: transitions are often found in the fossil record. Preserved transitions are not common—and should not be, according to our understanding of evolution (see next section)—but they are not entirely wanting, as creationists often claim. The lower jaw of reptiles contains several bones, that of mammals only one. The non-mammalian jawbones are reduced, step by step, in mammalian ancestors until they become tiny nubbins located at the back of the jaw. The "hammer" and "anvil" bones of the mammalian ear are descendants of these nubbins. How could such a transition be accomplished? the creationists ask. Surely a bone is either entirely in the jaw or in the ear. Yet paleontologists have discovered two transitional lineages of therapsids (the so-called mammal-like reptiles) with a double jaw joint—one composed of the old quadrate and articular bones (soon to become the hammer and anvil), the other of the squamosal and dentary bones (as in modern mammals). For that matter, what better transitional form could we expect to find than the oldest human, *Australopithecus afarensis,* with its apelike palate, its human upright stance, and a cranial capacity larger than any ape's of the same body size but a full 1,000 cubic centimeters below ours? If God made each of the half-dozen human species discovered in ancient rocks, why did he create in an unbroken temporal sequence of progressively more modern features—increasing cranial capacity, reduced face and teeth, larger body size? Did he create to mimic evolution and test our faith thereby?

Faced with these facts of evolution and the philosophical bankruptcy of their own position, creationists rely upon distortion and innuendo to buttress their rhetorical claim. If I sound sharp or bitter, indeed I am—for I have become a major target of these practices.

I count myself among the evolutionists who argue for a jerky, or episodic, rather than a smoothly gradual, pace of change. In 1972 my colleague Niles Eldredge and I developed the theory of punctuated equilibrium. We argued that two outstanding facts of the fossil record—geologically "sudden" origin of new species and failure to change thereafter (stasis)—reflect the predictions of evolutionary theory, not the imperfections of the fossil record. In most theories, small isolated populations are the source of new species, and the process of speciation takes thousands or tens of thousands of years. This amount of time, so long when measured against our lives, is a geological microsecond. It represents much less than 1 per cent of the average lifespan for a fossil invertebrate species—more than ten million years. Large, widespread, and well established species, on the other hand, are not expected to change very much. We believe that the inertia of large populations explains the stasis of most fossil species over millions of years.

We proposed the theory of punctuated equilibrium largely to provide a different explanation for pervasive trends in the fossil record. Trends, we argued, cannot be attributed to gradual transformation within lineages, but must arise from the differential success of certain kinds of species. A trend, we argued, is more like climbing a flight of stairs (punctuations and stasis) than rolling up an inclined plane.

Since we proposed punctuated equilibria to explain trends, it is infuriating to be quoted again and again by creationists—whether through design or stupidity, I do not know—as admitting that the fossil record includes no transitional forms. Transitional forms are generally lacking at the species level, but they are abundant between larger groups. Yet a pamphlet entitled "Harvard Scientists Agree Evolution Is a Hoax" states: "The facts of punctuated equilibrium which Gould and Eldredge . . . are forcing Darwinists to swallow fit the picture that Bryan insisted on, and which God has revealed to us in the Bible."

Continuing the distortion, several creationists have equated the theory of punctuated equilibrium with a caricature of the beliefs of Richard Goldschmidt, a great early geneticist. Goldschmidt argued, in a famous book published in 1940, that new groups can arise all at once through major mutations. He referred to these suddenly transformed creatures as "hopeful monsters." (I am attracted to some aspects of the non-caricatured version, but Goldschmidt's theory still has nothing to do with punctuated equilibrium—see essays in section 3 and my explicit essay on Goldschmidt in *The Panda's Thumb.*) Creationist Luther Sunderland talks of the "punctuated equilibrium hopeful monster theory" and tells his hopeful readers that "it amounts to tacit admission that anti-evolutionists are correct in asserting there is no fossil evidence supporting the theory that all life is connected to a common ancestor." Duane Gish writes, "According to Goldschmidt, and now apparently according to Gould, a reptile laid an egg from which the first bird, feathers and all, was produced." Any evolutionist who believed such nonsense would rightly be laughed off the intellectual stage; yet the only theory that could ever envision such a scenario for the origin of birds is creationism—with God acting in the egg.

I am both angry at and amused by the creationists; but mostly I am deeply sad. Sad for many reasons. Sad because so many people who respond to creationist appeals are troubled for the right reason, but venting their anger at the wrong target. It is true that scientists have often been dogmatic and elitist. It is true that we have often allowed the white-coated, advertising image to represent us—"Scientists say that Brand X cures bunions ten times faster than . . ." We have not fought it adequately because we derive benefits from appearing as a new priesthood. It is also true that faceless and bureaucratic state power intrudes more and more into our lives and removes choices that should belong to individuals and communities. I can understand that school curricula, imposed from above and without local input, might be seen as one more insult on all these grounds. But the culprit is not, and cannot be, evolution or any other fact of the natural world. Identify and fight your legitimate enemies by all means, but we are not among them.

I am sad because the practical result of this brouhaha will not be expanded coverage to include creationism (that would also make me sad), but the reduction or excision of evolution from high school curricula. Evolution is one of the half dozen "great ideas" developed by science. It speaks to the profound issues of genealogy that fascinate all of us—the "roots" phenomenon writ large. Where did we come from? Where did life arise? How did it develop? How are organisms related? It forces us to think, ponder, and wonder. Shall we deprive millions of this knowledge and once again teach biology as a set of dull and unconnected facts, without the thread that weaves diverse material into a supple unity?

But most of all I am saddened by a trend I am just beginning to discern among my colleagues. I sense that some now wish to mute the healthy debate about theory that has brought new life to evolutionary biology. It provides grist for creationist mills, they say, even if only by distortion. Perhaps we should lie low and rally round the flag of strict Darwinism, at least for the moment—a kind of old-time religion on our part.

But we should borrow another metaphor and recognize that we too have to tread a straight and narrow path, surrounded by roads to perdition. For if we ever begin to suppress our search to understand nature, to quench our own intellectual excitement in a misguided effort to present a united front where it does not and should not exist, then we are truly lost.

20 | A Visit to Dayton

IN HIS SUMMATION to the court, Clarence Darrow talked for three full days to save the lives of Nathan Leopold and Richard Loeb. Guilty they clearly were, of perhaps the most brutal and senseless murder of the 1920s. By arguing that they were victims of their upbringing, Darrow sought only to mitigate their personal responsibility and substitute a lifetime in jail for the noose. He won, as he usually did.

John Thomas Scopes, defendant in Darrow's next famous case, recalled his attorney's theory of human behavior in the opening lines of an autobiography published long after the famous "monkey trial" (see bibliography): "Clarence Darrow spent his life arguing teaching, really—that a man is the sum of his heredity and his environment." The world may seem capricious, but events have their reasons, however complex. These reasons conspire to drive events forward; Leopold and Loeb were not free agents when they bludgeoned Bobby Franks and stuffed his body into a culvert, all to test the idea that a perfect crime might be committed by men of sufficient intelligence.

We wish to find reasons for the manifest senselessness that surrounds us. But deterministic theories, like Darrow's, leave out the genuine randomness of our world, a chanciness that gives meaning to the old concept of human free will. Many events, although they move forward with accelerating inevitability after their inception, begin as a concatenation of staggering improbabilities. And so we all began, as one sperm among billions vying for entry; a microsecond later, I might have been the Stephanie my mother wanted.

The Scopes trial in Dayton, Tennessee, occurred as the outcome of accumulated improbability. The Butler Act, passed by the Tennessee legislature and signed by Gov. Austin Peay on March 21, 1925, declared it "unlawful for any teacher in any of the Universities, Normals and all other public schools of the state—which are supported in whole or in part by the public school funds of the State, to teach any theory that denies the story of the Divine Creation of man as taught in the Bible, and to teach instead that man has descended from a lower order of animals." The bill could have been beaten with little trouble had the opposition bothered to organize and lobby (as they had the previous year in Kentucky, when a similar bill in similar circumstances went down to easy defeat). The senate passed it with no enthusiasm, assuming a gubernatorial veto. One member said of Mr. Butler: "The gentleman from Macon wanted a bill passed; he had not had much during the session and this did not amount to a row of pins; let him have it." But Peay, admitting the bill's absurdity and protesting that the legislature should have saved him from embarrassment by defeating it, signed the act as an innocuous statement of Christian principles: "After a careful examination," wrote Peay, "I can find nothing of consequence in the books now being taught in our schools with which this bill will interfere in the slightest manner. Therefore it will not put our teachers in any jeopardy. Probably the law will never be applied. . . . Nobody believes that it is going to be an active statute." (See Ray Ginger's *Six Days or Forever?* for a fine account of the legislative debate.)

If the bill itself was improbable, Scopes's test of it was even more unlikely. The American Civil Liberties Union (ACLU) offered to supply council and provide legal costs for any teacher willing to challenge the act by courting an arrest for teaching evolution. The test was set for the favorable urban setting of Chattanooga, but plans fell through.

Scopes didn't even teach biology in the small, inappropriate, fundamentalist town of Dayton, located forty miles north of Chattanooga. He had been hired as an athletic coach and physics teacher but had substituted in biology when the regular instructor (and principal of the school) fell ill. He had not actively taught evolution at all, but merely assigned the offending textbook pages as part of a review for an exam. When some town boosters decided that a test of the Butler Act might put Dayton on the map—none showed much interest in the intellectual issues—Scopes was available only by another quirk of fate. (They would not have asked the principal, an older, conservative family man, but they suspected that Scopes, a bachelor and free thinker, might go along.) The school year was over, and Scopes had intended to depart immediately for a summer with his family. But he stayed on because he had a date with "a beautiful blonde" at a forthcoming church social.

Scopes was playing tennis on a warm afternoon in May, when a small boy appeared with a message from "Doc" Robinson, the local pharmacist and owner of Dayton's social center, Robinson's Drug Store. Scopes finished his game, for there is no urgency in Dayton, and then ambled on down to Robinson's, where he found Dayton's leading citizens crowded around a table, sipping Coke and arguing about the Butler Act. Within a few minutes, Scopes had offered himself as the sacrificial lamb. From that point, events accelerated and began to run along a predictable track. William Jennings Bryan, who had stirred millions with his "Cross of Gold" speech and almost become president as a result, was passing his declining years as a fundamentalist stumper—"a tinpot pope in the Coca-Cola belt," as H. L. Mencken remarked. He volunteered his services for the prosecution, and Clarence Darrow responded in kind for the defense. The rest, as they say, is history. Of late, it has, alas, become current events as well.

Robinson's Drug Store is still the social center of Dayton, although it moved in 1928 to its present location in the shadow of the Rhea County courthouse, where Scopes faced the wrath of Bryan's God. "Sonny" Robinson, Doc's

Robinson's Drug Store, where it all began; moved to its present location in the late 1920s. PHOTO BY DEBORAH GOULD.

boy, has run the store for decades, dispensing pills to the local citizenry and thoughts about Dayton's moment of fame to the pilgrims and gawkers who stop by to see where it all started. The little round table, with its wire-backed chairs, occupies a central place, as it did when Scopes, Doc Robinson, and George Rappelyea (who made the formal "arrest") laid their plans in May 1925. The walls are covered with pictures and other memorabilia, including Sonny Robinson's only personal memory of the trial: a photograph of a five-year-old boy, sitting in a carriage and pouting because a chimpanzee had received the Coke he had expected. (The chimp was a prominent member in the motley entourage of camp followers, many of comparable intelligence, that descended upon Dayton during the trial, in search of ready cash rather than eternal enlightenment.)

I was visiting Robinson's Drug Store in June 1981, when a San Francisco paper called with a request for photos of modern Dayton. Sonny Robinson, who claims to be a shy

The interior of Robinson's is covered with photos and other mementos of the Scopes trial. PHOTO BY DEBORAH GOULD.

man, began a flurry of calls to exploit the moment. Up north in the big town, you wouldn't keep a man waiting, at least not without a request or an explanation: "Excuse me, I know you must be in a hurry, but would you mind, it won't be more than a few minutes. . . ." But it was 97° outside and cool in Sonny Robinson's store. And where would a man be going anyway? Half an hour later, his personages assembled, Sonny Robinson pulled out the famous table and brought three Cokes in some old-fashioned five-cent glasses. I sat in the middle ("the biology professor from Harvard who just happened to walk in," as Robinson had told his callers). On one side sat Ted Mercer, president of Bryan College, the fundamentalist school begun as a legacy to the "Great Commoner's" last battle. On the other side sat Mr. Robinson, son of the man who had started it all around the same table fifty-six years before. The fundamentalist editor of the Dayton *Herald* snapped our pictures and we sipped our Cokes.

Dayton has remained a small and inconspicuous town. If you're coming from Knoxville via Decatur, you still have to cross the Tennessee River on a six-car ferry. The older

A proper way to treat the issues that divide us. Three men of divergent views chat and sip Coke aound the "original" table in Robinson's Drug Store. Left, Ted Mercer, president of fundamentalist Bryan College; center, yours truly; right, Sonny Robinson. PHOTO BY DEBORAH GOULD.

houses are well kept, with four white pillars in front, the vernacular imitation of plantation style. (As a regional marker of the South, these pillars are architecture's equivalent of the dependable gastronomical criterion: when the beverage simply labeled "tea" on the menu invariably comes iced.) H. L. Mencken, not known for words of praise, confessed (in surprise) his liking for Dayton:

> I had expected to find a squalid Southern village . . . with pigs rooting under the houses and the inhabitants full of hookworm and malaria. What I found was a country town full of charm and even beauty. . . . The

> houses are surrounded by pretty gardens, with cool green lawns and stately trees. . . . The stores carry good stocks and have a metropolitan air, especially the drug, book, magazine, sporting goods and soda-water emporium of the estimable Robinson.

Some things have changed, of course. Trailers now rooted to their turf and houses of undressed concrete block reflect the doubling of Dayton's population to nearly 4,000. The older certainties may have eroded somewhat. A banner headline in this week's Dayton *Herald* tells of a $200-million marijuana crop confiscated and destroyed in Rhea and neighboring Bledsoe counties. And a quarter gets you a condom—"sold for the prevention of disease only," of course—at vending machines in restrooms of local service stations. At least they can't blame evolution for this, as one evangelical minister did a few months back when he cited Darwin as a primary supporter of the four "p's": prostitu-

The Rhea County Court House in Dayton, Tennessee, scene of the Scopes trial. PHOTO BY DEBORAH GOULD.

tion, perversion, pornography, and permissiveness. They taught creationism in Dayton before John Scopes arrived, and they teach it today.

For all these muted changes, Dayton remains a two-street town, dwarfed at the crossroad by the Rhea County courthouse, a Renaissance Revival building of the 1890s seemingly too large by half for a small town in a small county. Yet even this courtroom failed in its moment of glory, as Judge Raulston, noting that the weight of humanity had opened cracks in the ceiling below, reconvened his court on the side lawn, where Darrow grilled Bryan alfresco. (It is a meaningless and tangential irony to be sure, but I thought I'd mention it. Rhea was the daughter of Uranus and the mother of Zeus. Her name also applies to the South American "ostrich." On the *Beagle* voyage, Darwin rediscovered a second species, the lesser, or Darwin's, rhea, living in a different part of South America. In one of his first evolutionary speculations, Darwin surmised that the spatial difference between these two rheas might be analogous to the temporal distinction between extinct species and their living relatives.)

The Scopes trial is surrounded by misconceptions, and their exposure provides as good a way as any for recounting the basic story. In the heroic version, John Scopes was persecuted, Darrow rose to Scopes's defense and smote the antediluvian Bryan, and the antievolution movement then dwindled or ground to at least a temporary halt. All three parts of this story are false.

For the first, we have already noted that Austin Peay and the legislators of Tennessee did not intend to enforce their law. In fact, Bryan himself had lobbied the legislature (unsuccessfully) with advice that the act prescribe no penalty for noncompliance. It was, after all, only a symbolic statement; any teeth in the act might lead to its upset on constitutional grounds. The ACLU advertised in Tennessee papers for a test case. George Rappelyea read their offer in the Chattanooga *Times* and moseyed on down to Robinson's with his plan. Later, John Scopes remembered: "It was just a drugstore discussion that got past control." For once, the tale of outside (even Yankee) agitators tells at least a half-truth.

Bryan was vanquished and embarrassed during the trial, but not primarily by Darrow. The trial itself, with its foregone conclusion, was something of a bore. (Scopes *had* violated the law and the defense wanted a quick conviction for an advantageous move to a higher court.) It dragged on in interminable legal wrangling and had only two moments of high drama. The first occurred during a legal argument about the admissibility of expert testimony. The defense had brought to Dayton an impressive array of men prominent both in evolutionary biology and Christian conviction. The prosecution urged that their testimony be excluded. The law plainly forbade the teaching of human evolution, and Scopes just as plainly had violated the law. The potential truth of evolution was not at issue. Judge Raulston agreed with the prosecution, and the assembled experts took to their typewriters instead. With benefit of a weekend recess to hone their statements, the experts produced some formidable documents. They were printed in newspapers throughout the country, and Judge Raulston finally did admit them into the printed record of the trial.

Bryan, who had sat in uncharacteristic silence for several days, used this procedural argument as a springboard for his prepared excoriation of evolution. In a grandiloquent speech clearly directed to his constituents (witnesses remarked that he stood with his back to the judge), Bryan virtually denied that humans were mammals and argued that the case of Messrs. Leopold and Loeb amply demonstrated that too much learning is a dangerous thing. The defense's rebuttal provided Bryan's first humiliation; for in this rural land, before the advent of television, no art commanded more respect than speechifying. And Bryan was just plain outspoken, outgestured, and outshouted—not by Darrow, but by another defense attorney, Dudley Field Malone, a prominent New York divorce lawyer and former subordinate to Bryan in the State Department (where Bryan had been secretary under Woodrow Wilson). H. L. Mencken wrote:

> I doubt that any louder speech has ever been heard in a court of law since the days of Gog and Magog. It

> roared out of the open windows like the sound of artillery practice, and alarmed the moonshiners and catamounts on distant peaks. . . . In brief, Malone was in good voice. It was a great day for Ireland. And for the defense. For Malone not only out-yelled Bryan, he also plainly out-generaled and out-argued him. . . . It conquered even the fundamentalists. At its end they gave it a tremendous cheer—a cheer at least four times as hearty as that given to Bryan. For these rustics delight in speechifying, and know when it is good. The devil's logic cannot fetch them, but they are not above taking a voluptuous pleasure in his lascivious phrases.

Nonetheless, Judge Raulston ruled against the defense and excluded expert testimony the next morning. It was Friday and all seemed over, including the shouting. Scopes would be convicted summarily on Monday morning; he had violated the law and Raulston's narrow construction of the case had excluded all other issues. Virtually all the journalists, including H. L. Mencken, left town to avoid both the lull of a weekend recess and the expected anticlimax to follow. Thus, when Darrow induced Bryan to take the stand as an expert witness on the Bible, he spoke to a depleted local crowd and a skeleton crew of journalists. The reconstructions of *Inherit the Wind* and other accounts dramatize what was only an afterthought.

It is not even clear why Raulston allowed Bryan to appear (since he had excluded experts of opposite persuasion). The other prosecuting attorneys tried to dissuade Bryan, and Raulston finally expunged the entire exchange from the record. Bryan viewed the occasion as a desperate attempt to recover from Malone's drubbing, but Darrow exposed him as a pompous fool. Still, the most famous moment of the exchange—when Bryan deserted strict fundamentalist tenets by admitting that the days of Genesis might have lasted far longer than twenty-four hours—was not, as legend has it, a reluctant admission drawn forth by Darrow's ruthless logic. Bryan offered this statement freely, as an initial response to a series of questions. He did not seem to appreciate that local fundamentalists would regard it as a betrayal, and the surrounding world as a fatal inconsistency.

Scope's conviction was eventually quashed on a technicality. Judge Raulston had set the fine of $100 himself, but Tennessee law required that all fines greater than $50 be recommended by the jury. With Bryan humiliated and the conviction quashed, the legend of victory for the defense arose, thus completing the heroic version. But the Scopes trial was a defeat (or a victory so Pyrrhic that it scarcely deserves the name) for several reasons. First, Bryan recouped by involuntarily taking the only option left for an immediate restoration of prestige: he died in Dayton a week after the trial ended. Ted Mercer's Bryan College, a thriving fundamentalist institution in Dayton, is his local legacy. Second, the quashing of Scopes's conviction was a bitter pill for the defense. Suddenly, there was no case left to appeal. All that effort down the tubes of a judge's $50 error.

The Butler Act remained on the books until its repeal in 1967. It was not enforced, but who can tell how many teachers muted or suppressed their views and how many children never learned one of the most exciting and expansive ideas ever developed by scientists. In 1973, a "Genesis Bill" passed the senate of Tennessee, 69 to 16. It legislated equal time for evolution and creation and required a disclaimer in all texts that any stated idea about "the origin and creation of man and his world . . . is not represented to be scientific fact." The Bible, however, was declared a reference work, not a text, and therefore exempt from the requirement for a printed disclaimer. This bill was declared unconstitutional a few years later.

Third, and sadly, any hope that the issues of Scopes's trial had been banished to the realm of nostalgic Americana have been swept aside by our current creationist resurgence—the climate that inspired my own detour across the Tennessee River.

Late in his life, I came to know Kirtley Mather, emeritus professor of geology at Harvard, pillar of the Baptist church, lonely defender of academic freedom during the worst days of McCarthyism, and perhaps the finest man I have ever known. Kirtley was also a defense witness in Dayton. Each year, from the late 1960s to the mid 1970s, Kirtley gave a lecture to my class recalling his experiences at Dayton. It seemed a wonderful echo of times gone by, for Kirtley, in his late eighties, could still weave circles around the finest orators at Harvard. The lecture didn't change much from year to year. I viewed it first as a charming evocation, later as mildly related to current affairs, finally as a vital statement of pressing realities. This year, I will dust off the videotape and show it to my class as a disquisition on immediate dangers.

In 1965, John Scopes permitted himself this hope in retrospect:

> I believe that the Dayton trial marked the beginning of the decline of fundamentalism. . . . I feel that restrictive legislation on academic freedom is forever a thing of the past, that religion and science may now address one another in an atmosphere of mutual respect and of a common quest for truth. I like to think that the Dayton trial had some part in bringing to birth this new era.

(Scopes, by the way, later went to the University of Chicago and became a geologist. He lived quietly in Shreveport, Louisiana, for most of his life, working, as do so many geologists, for the petroleum industry. This splendid man of quiet integrity refused to capitalize on his transient and accidental fame in any way. His silence betokened no crisis of confidence or any departure from the principles that led to his momentary renown. He simply chose to make his own way on his own merits.)

Today, Jerry Falwell has donned Bryan's mantle, and Scopes's hopes for a "new era" have been thwarted. Of course, we will not replay Scopes's drama in exactly the same way; we have advanced somewhere in fifty-six years. Evolution is now too strong to exclude entirely, and current proposals for legislation mandate "equal time" for evolution and for old-time religion masquerading under the self-contradictory title of "scientific creationism." But the similarities between 1925 and 1981 are more disconcerting than the differences are comforting.

As in 1925, creationists are not battling for religion. They have been disowned by leading churchmen of all persuasions, for they debase religion even more than they misconstrue science. They are a motley collection to be sure, but their core of practical support lies with the evangelical right, and creationism is a mere stalking horse or subsidiary issue in a political program that would ban abortion, erase the political and social gains of women by reducing the vital concept of the family to an outmoded paternalism, and reinstitute all the jingoism and distrust of learning that prepares a nation for demagoguery.

As in 1925, they use the same methods of willful misquotation to impart a "scientific" patina to creationism. I am now a major victim of these efforts because my views on rapid evolutionary bursts followed by long periods of stasis can be distorted to apparent support for creation by fiat and unchanging persistence of immutable types (see last essay). I was therefore amused (or soothed) to read that, in 1925, Bryan and company were using the same strategy to exploit the forthright address that William Bateson had delivered to the American Association for the Advancement of Science in 1922. Bateson had expressed his confidence in the fact of evolution, but had honestly admitted that, despite reigning pomposity and textbook pap, we knew rather little about the mechanisms of evolutionary change. Scopes's prosecutors had cited Bateson as "proof" that scientists had admitted the tenuousness of evolution itself. In his written affidavit to Judge Raulston, W. C. Curtis, a zoologist from the University of Missouri, submitted a letter from Bateson:

> I have looked through my Toronto address again. I see nothing in it which can be construed as expressing doubt as to the main fact of evolution. . . . I took occasion to call the attention of my colleagues to the loose thinking and unproven assumptions which pass current as to the actual processes of evolution. We do

> know that the plants and animals, including most certainly man, have been evolved from other and very different forms of life. As to the nature of this process of evolution, we have many conjectures but little positive knowledge.

Curtis also submitted a letter from Bryan's old boss Woodrow Wilson: "Of course, like every other man of intelligence and education, I do believe in organic evolution. It surprises me that at this late date such questions should be raised." Ronald Reagan, however, raised them on the stump before a fundamentalist crowd in Dallas.

I learned something in Dayton, or rather, remembered something I should never have forgotten, while sipping that five-cent Coke around the original table in Robinson's Drug Store. The enemy is not fundamentalism; it is intolerance. In this case, the intolerance is perverse since it masquerades under the "liberal" rhetoric of "equal time." But mistake it not. Creationists are trying to impose a specific religious view by legislative fiat upon teachers who reject it both by conscience and training. For all their talk about weighing both sides (a mere question of political expediency), they would also substitute biblical authority for free scientific inquiry as a source of empirical knowledge.

All the commentators at Dayton, including the caustic H. L. Mencken himself, noted that the local people, although secure in their creationist beliefs, showed no intolerance or even discourtesy to the opposition. They feted Bryan when he arrived in town, and they provided a spread of equal size for Darrow. They applauded Malone on the merits of his speech. Mencken wrote:

> Nor is there any evidence in the town of that poisonous spirit which usually shows itself when Christian men gather to defend the great doctrine of their faith. I have heard absolutely no whisper that Scopes is in the pay of the Jesuits, or that the whisky trust is backing him, or that he is egged on by the Jews who manufacture

> lascivious moving pictures. On the contrary, the Evolutionists and the Anti-Evolutionists seem to be on the best of terms and it is hard in a group to distinguish one from the other.

Dayton has persevered in its geniality. I encountered warm disagreement with my evolutionary views, but I sensed no disrespect for my opinions and no inclination to demote me as a person because I disagreed with a favored belief. Geniality of this sort is widespread but, alas, so fragile. A few seeds of ugliness and intolerance can generate the cover for an entire field. We have nothing to fear from the vast majority of fundamentalists who, like many citizens of Dayton, live by a doctrine that is legitimately indigenous to their area. Rather, we must combat the few yahoos who exploit the fruits of poor education for ready cash and larger political ends.

Bryan was easy prey for ridicule, since he was such a fool in his political dotage. Mencken wrote in his sharpest prose:

> Once he had one leg in the White House and the nation trembled under his roars. Now he is a tinpot pope in the Coca-Cola belt and a brother to the forlorn pastors who belabor half-wits in galvanized iron tabernacles behind the railroad yards. . . . It is a tragedy, indeed, to begin life as a hero and to end it as a buffoon.

Many current creationists seem equally pitiable in their pronouncements: can anyone take seriously a link between Darwinism and the four evil p's?

But Mencken also understood the dangers, for he wrote in his final lines:

> Let no one mistake it for comedy, farcical though it may be in all its details. It serves notice on the country that Neanderthal man is organizing in these forlorn backwaters of the land, led by a fanatic, rid of sense and devoid of conscience. Tennessee, challenging him too timor-

ously and too late, now sees its courts converted into camp meetings and its Bill of Rights made a mock of by its sworn officers of the law.

Do movements of intolerance ever start in any other way, given our pervasive tendencies toward geniality? Do they not always begin in comedy and end, when successful, in carnage? Who did not regard Hitler as an object of pitiful derision after the beer hall putsch. And who can read the famous words of Protestant theologian Martin Niemöller without a shudder:

> First the Nazis went after the Jews, but I wasn't a Jew, so I did not react. Then they went after the Catholics, but I wasn't a Catholic, so I didn't object. Then they went after the workers, so I didn't stand up. Then they went after the Protestant clergy and by then it was too late for anybody to stand up.

Clarence Darrow understood the roots of intolerance only too well when he said in Dayton:

> If today you can take a thing like evolution and make it a crime to teach it in the public schools, tomorrow you can make it a crime to teach it in the private schools and next year you can make it a crime to teach it to the hustings or in the church. At the next session you may ban books and the newspapers. . . . Ignorance and fanaticism are ever busy and need feeding. Always feeding and gloating for more. Today it is the public school teachers; tomorrow the private. The next day the preachers and the lecturers, the magazines, the books, the newspapers. After a while, Your Honor, it is the setting of man against man and creed against creed until with flying banners and beating drums we are marching backward to the glorious ages of the sixteenth century when bigots lighted fagots to burn the men who dared to bring any intelligence and enlightenment and culture to the human mind.

Ever the cynic, H. L. Mencken evaluated this impassioned plea: "The net effect of Clarence Darrow's great speech yesterday seems to be precisely the same as if he had bawled it up a rainspout in the interior of Afghanistan." We had better proclaim the same message today into a downspout that resonates across the nation.

21 | Moon, Mann, and Otto

Little Rock, Arkansas
December 10, 1981

THIS MORNING'S *Arkansas Gazette* features a cartoon with searchlights focused on a state map. The map displays neither topography nor political boundaries, but merely contains the words, etched in black from Oklahoma to the Mississippi: "Scopes Trial II. Notoriety." I spent most of yesterday—with varying degrees of pleasure, righteousness, discomfort, and disbelief—in the witness box, trying to convince Federal Judge William R. Overton that all the geological strata on earth did not form as the result of a single Noachian deluge. We are engaged in the first legal test upon the new wave of creationist bills that mandate equal time or "balanced treatment" for evolution and a thinly disguised version of the Book of Genesis read literally, but masquerading under the nonsense phrase "creation science." The judge, to say the least, seems receptive to my message and as bemused as I am by the fact that such a trial can be held just a few months before the hundredth anniversary of Darwin's death.

The trial of John Scopes in 1925 has cast such a long shadow into our own times that the proceedings in Little Rock inevitably invite comparison (see last essay). I appreciate the historical continuity but am more impressed by the differences. I sit in a massive alabaster building, a combined courthouse and post office, a no-nonsense, no-frills edifice, surrounded by the traffic noises of downtown Little Rock. The Rhea County Courthouse of Dayton, Tennessee—the building that hosted Scopes, Darrow, and Bryan in 1925—is a gracious, shaded, and decorated Renaissance Revival structure that dominates the crossroads of its two-street town. The Scopes trial was directly initiated by Dayton's boosters to put their little town on the map; many, probably most, citizens of Arkansas are embarrassed by the anachronism on their doorstep. John Scopes was convicted for even mentioning that humans had descended from "a lower order of animals"; we have made some progress in half a century, and modern creationists clamor for the official recognition of their pseudoscience, not (at least yet) for the exclusion of our well-documented conclusions.

I decided to be a paleontologist when I was five, after an awestruck encounter with *Tyrannosaurus* at the Museum of Natural History in New York. The phenomenology of big beasts might have been enough to sustain my interest, but I confirmed my career six years later when I read, far too early and with dim understanding, G. G. Simpson's *Meaning of Evolution* and discovered that a body of exciting ideas made sense of all those bodies of bone. Three years later, I therefore approached my first high school science course with keen anticipation. In a year of biology, I would surely learn all about evolution. Imagine my disappointment when the teacher granted Mr. Darwin and his entire legacy only an apologetic two days at the very end of a trying year. I always wondered why, but was too shy to ask. Then I just forgot my question and continued to study on my own.

Six months ago, in a secondhand bookstore, I found a copy of my old high school text, *Modern Biology,* by T. J. Moon, P. B. Mann, and J. H. Otto. We all appreciate how powerful an unexpected sight or odor can be in triggering a distant "remembrance of things past." I knew what I had the minute I saw that familiar red binding with its embossed microscope in silver and its frontispiece in garish color, showing a busy beaver at work. The book, previously the property of a certain "Lefty," was soon mine for ninety-five cents.

Now, more than half a life later (I studied high school biology in 1956), I finally understand why Mrs. Blenderman had neglected the subject that so passionately interested me. I had been a victim of Scopes's ghost (or rather, of his adversary, Bryan's). Most people view the Scopes trial as a victory for evolution, if only because Paul Muni and Spencer Tracy served Clarence Darrow so well in theatrical and film versions of *Inherit the Wind,* and because the trial triggered an outpouring of popular literature by aggrieved and outraged evolutionists. Scopes's conviction (later quashed on a technicality) had been a mere formality; the battle for evolution had been won in the court of public opinion. Would it were so. As several historians have shown, the Scopes trial was a rousing defeat. It abetted a growing fundamentalist movement and led directly to the dilution or elimination of evolution from all popular high school texts in the United States (see bibliography for works of Grabiner and Miller, and of Nelkin). No arm of the industry is as cowardly and conservative as the publishers of public school texts—markets of millions are not easily ignored. The situation did not change until 1957, a year too late for me, when the Russian Sputnik provoked a searching inquiry into the shameful state of science education in America's high schools.

Moon, Mann, and Otto commanded the lion's share of the market in the mid-1950s; readers of my generation will probably experience that exhilarating sense of *déjà vu* with me. Like many popular books, it was the altered descendant of several earlier editions. The first, *Biology for Beginners,* by Truman J. Moon, was published in 1921, before the Scopes trial. Its frontispiece substituted Mr. Darwin for the industrious beaver, and its text reflected a thorough immersion in evolution as the focal subject of the life sciences. Its preface proclaimed: "The course emphasizes the fact that biology is a unit science, based on the fundamental idea of evolution rather than a forced combination of portions of botany, zoology and hygiene." Its text contains several chapters on evolution and continually emphasizes Darwin's central contention that the *fact* of evolution is established beyond reasonable doubt, although scientists have much to learn about the *mechanism* of evolutionary change (see essay 19). Chapter 35, on "The Method of Evolution," begins: "Proof of the *fact* of similarity between the various forms of living things and of their very evident relationship, still leaves a more difficult question to be answered. *How* did this descent and modification take place, by what means has nature developed one form from another? [Moon's italics]"

I then examined my new purchase with a growing sense of amusement mixed with disgust. The index contained such important entries as "fly specks, disease germs in," but nothing about evolution. Indeed, the word evolution does not occur anywhere in the book. The subject is not, however, entirely absent. It receives a scant eighteen pages in a 662-page book, as chapter 58 of 60 (pp. 618–36). In this bowdlerized jiffy, it is called "The hypothesis of racial development." Moon, Mann, and Otto had gone the post-Scopes way of all profitable texts: eliminate and risk no offense. (Those who recall the reality of high school courses will also remember that many teachers never got to those last few chapters at all.)

This one pussyfooting chapter is as disgraceful in content as in brevity. Its opening two paragraphs are a giveaway and an intellectual sham compared with Moon's forthright words of 1921. The first paragraph provides a fine statement of historical continuity and change in the *physical* features of our planet:

> This is a changing world. It changes from day to day, year to year, and from age to age. Rivers deepen their gorges as they carry more land to the sea. Mountains rise, only to be leveled gradually by winds and rain. Continents rise and sink into the sea. Such are the gradual changes of the physical earth as days add into years and years combine to become ages.

Now what could be more natural and logical than to extend this same mode of reasoning and style of language to life? The paragraph seems to be set up for such a transition. But note how the tone of the second paragraph subtly shifts to avoid any commitment to historical continuity for organic change:

> During these ages, species of plants and animals have appeared, have flourished for a time, and then have perished as new species took their places. . . . When one race lost in the struggle for survival, another race appeared to take its place.

Four pages later, we finally get an inkling that genealogy may be behind organic transitions through time: "This geological story of the rocks, showing fossil gradations from simple to complex organisms, is what we should expect to find if there had been racial development throughout the past." Later on the page, Moon, Mann, and Otto ask the dreaded question and even venture the closest word they dare to "evolution": "Are these prehistoric creatures the ancestors of modern animals?" If you read carefully through all the qualifications, they answer their question with a guarded "yes"—but you have to read awfully hard.

Thus were millions of children deprived of their chance to study one of the most exciting and influential ideas in science, the central theme of all biology. A few hundred, myself included, possessed the internal motivation to transcend this mockery of education, but citing us seems as foolish and cruel as the old racist argument, "what about George Washington Carver or Willie Mays," used to refute the claim that poor achievement might be linked to economic disadvantage and social prejudice.

Now I can mouth all the grandiloquent arguments against such a dilution of education: we will train a generation unable to think for themselves, we will weaken the economic and social fabric of the nation if we raise a generation illiterate in science, and so on. I even believe all these arguments. But this is not what troubled me most as I read chapter 58 in Moon, Mann, and Otto. I wasn't even much angered, but merely amused, by the tortured pussyfooting and glaring omissions. Small items with big implications are my bread and butter, as any reader of these essays will soon discover. I do not react strongly to generalities. I can ignore a displeasing general tenor, but I cannot bear falsification and debasement of something small and noble. I was not really shaken until I read the last paragraph of chapter 58, but then an interior voice rose up and began to compose this essay. For to make a valid point in the context of their cowardice, Moon, Mann, and Otto had perverted (perhaps unknowingly) one of my favorite quotations. If cowardice can inspire such debasement, then it must be rooted out.

The last paragraph is titled: Science and Religion. I agree entirely with its first two sentences: "There is nothing in science which is opposed to a belief in God and religion. Those who think so are mistaken in their science or their theology or both." They then quote (with some minor errors, here corrected) a famous statement of T. H. Huxley, using it to argue that a man may be both a Darwinian and a devout Christian:

> Science seems to me to teach in the highest and strongest manner the great truth which is embodied in the Christian conception of entire surrender to the will of God. Sit down before fact as a little child, be prepared to give up every preconceived notion, follow humbly wherever and to whatever abysses nature leads, or you shall learn nothing. I have only begun to learn content and peace of mind since I have resolved at all risks to do this.

Now a man may be both an evolutionist and a devout Christian. Millions successfully juxtapose these two independent viewpoints, but Thomas Henry Huxley did not. This quote, in its proper context, actually speaks of Huxley's courageous agnosticism. It also occurs in what I regard as the most beautiful and moving letter ever written by a scientist.

The tragic setting of this long letter explains why Huxley cited, only in analogy as Moon, Mann, and Otto did not understand, "the Christian conception of entire surrender to the will of God." Huxley's young and favorite son had just died. His friend, the Reverend Charles Kingsley (best remembered today as author of *The Water-Babies* and *Westward Ho!*) had written a long and kind letter of condolence with a good Anglican bottom line: see here Huxley, if you could only abandon your blasted agnosticism and accept the Christian concept of an immortal soul, you would be comforted.

Huxley responded in tones that recall the chief of police in Gilbert and Sullivan's *Pirates of Penzance* who, when praised by General Stanley's daughters for expected bravery in a coming battle that would probably lead to his bloody death, remarked:

> Still, perhaps it would be wise
> Not to carp or criticise,
> For it's very evident
> These attentions are well meant.

Huxley thanks Kingsley for his sincerely proffered comfort, but then explains in several pages of passionate prose why he cannot alter a set of principles, established after so much thought and deliberation, merely to assuage his current grief.

He has, he maintains, committed himself to science as the only sure guide to truth about matters of fact. Since matters of God and soul do not lie in this realm, he cannot know the answers to specific claims and must remain agnostic. "I neither deny nor affirm the immortality of man," he writes. "I see no reason for believing in it, but, on the other hand, I have no means of disproving it." Thus, he continues, I cannot assert the certainty of immortality to placate my loss. Uncomfortable convictions, if well founded, are those that require the most assiduous affirmation, as he states just before the passage quoted by Moon, Mann, and Otto: "My business is to teach my aspirations to conform themselves to fact, not to try and make facts harmonize with my aspirations."

Later, in the most moving statement of the letter, he speaks of the larger comfort that a commitment to science has provided him—a comfort more profound and lasting than the grief that his uncertainty about immortality now inspires. Among three agencies that shaped his deepest beliefs, he notes, "Science and her methods gave me a resting-place independent of authority and tradition." (For his two other agencies, Huxley cites "love" that "opened up to me a view of the sanctity of human nature," and his recognition that "a deep sense of religion was compatible with the entire absence of theology.") He then writes:

> If at this moment I am not a worn-out, debauched, useless carcass of a man, if it has been or will be my fate to advance the cause of science, if I feel that I have a shadow of a claim on the love of those about me, if in the supreme moment when I looked down into my boy's grave my sorrow was full of submission and without bitterness, it is because these agencies have worked upon me, and not because I have ever cared whether my poor personality shall remain distinct forever from the All from whence it came and whither it goes.
>
> And thus, my dear Kingsley, you will understand what my position is. I may be quite wrong, and in that case I know I shall have to pay the penalty for being wrong. But I can only say with Luther, "Gott helfe mir, ich kann nichts anders [God help me, I cannot do otherwise]."

Thus we understand what Huxley meant when he spoke of "the Christian conception of entire surrender to the will of God" in the passage cited by Moon, Mann, and Otto. It is obviously not, as they imply, his profession of Christian faith, but a burning analogy: as the Christian has made his commitment, so have I made mine to science. I cannot do otherwise, despite the immediate comfort that conventional Christianity would supply in my current distress.

Today I sat in the court of Little Rock, listening to the testimony of four splendid men and women who teach science in primary and secondary schools of Arkansas. Their testimony contained moments of humor, as when one teacher described an exercise he uses in the second grade. He stretches a string across the classroom to represent the age of the earth. He then asks students to stand in various positions marking such events as the origin of life, the extinction of dinosaurs, and the evolution of humans. What would you do, asked the assistant attorney general in cross-examination, to provide balanced treatment for the 10,000-year-old earth advocated by creation scientists. "I guess I'd have to get a short string," replied the teacher. The thought of twenty earnest second graders, all scrunched up along a millimeter of string, created a visual image that set the court rocking with laughter.

But the teachers' testimony also contained moments of inspiration. As I listened to their reasons for opposing "creation science," I thought of T. H. Huxley and the courage required by dedicated people who will not, to paraphrase Lillian Hellman, tailor their convictions to fit current fashions. As Huxley would not simplify and debase in order to find immediate comfort, these teachers told the court that mechanical compliance with the "balanced treatment" act, although easy enough to perform, would destroy their integrity as teachers and violate their responsibility to students.

One witness pointed to a passage in his chemistry text that attributed great age to fossil fuels. Since the Arkansas act specifically includes "a relatively recent age of the earth" among the definitions of creation science requiring "balanced treatment," this passage would have to be changed. The witness claimed that he did not know how to make such an alteration. Why not? retorted the assistant attorney general in his cross-examination. You only need to insert a simple sentence: "Some scientists, however, believe that fossil fuels are relatively young." Then, in the most impressive statement of the entire trial, the teacher responded. I could, he argued, insert such a sentence in mechanical compliance with the act. But I cannot, as a conscientious teacher, do so. For "balanced treatment" must mean "equal dignity" and I would therefore have to justify the insertion. And this I cannot do, for I have heard no valid arguments that would support such a position.

Another teacher spoke of similar dilemmas in providing balanced treatment in a conscientious rather than a mechanical way. What then, he was asked, would he do if the law were upheld. He looked up and said, in his calm and dignified voice: It would be my tendency not to comply. I am not a revolutionary or a martyr, but I have responsibilities to my students, and I cannot forego them.

God bless the dedicated teachers of this world. We who work in unthreatened private colleges and universities often do not adequately appreciate the plight of our colleagues—or their courage in upholding what should be our common goals. What Moon, Mann, and Otto did to Huxley epitomizes the greatest danger of imposed antirationalism in classrooms—that one must simplify by distortion, and remove both depth and beauty, in order to comply.

In appreciation for the teachers of Arkansas, then, and for all of us, one more statement in conclusion from Huxley's letter to Kingsley:

> Had I lived a couple of centuries earlier I could have fancied a devil scoffing at me . . . and asking me what profit it was to have stripped myself of the hopes and consolations of the mass of mankind? To which my only reply was and is—Oh devil! truth is better than much profit. I have searched over the grounds of my belief, and if wife and child and name and fame were all to be lost to me one after the other as the penalty, still I will not lie.

Postscript

On January 5, 1982, Federal District Judge William R. Overton declared the Arkansas act unconstitutional because it forces biology teachers to purvey religion in science classrooms.

22 Science and Jewish Immigration

IN APRIL 1925, C. B. Davenport, one of America's leading geneticists, wrote to Madison Grant, author of *The Passing of the Great Race,* and the most notorious American racist of the genteel Yankee tradition: "Our ancestors drove Baptists from Massachusetts Bay into Rhode Island, but we have no place to drive the Jews to." If America had become too full to provide places of insulated storage for undesirables, then they must be kept out. Davenport had written Grant to discuss a pressing political problem of the day: the establishment of quotas for immigration to America.

Jews presented a potential problem to ardent restrictionists. After 1890, the character of American immigration had changed markedly. The congenial Englishmen, Germans, and Scandinavians, who predominated before, had been replaced by hordes of poorer, darker, and more unfamiliar people from southern and eastern Europe. The catalog of national stereotypes proclaimed that all these people—primarily Italians, Greeks, Turks, and Slavs—were innately deficient in both intelligence and morality. Arguments for exclusion could be grounded in the eugenic preservation of a threatened American stock. But Jews presented a dilemma. The same racist catalog attributed a number of undesirable traits to them, including avarice and inability to assimilate, but it did not accuse them of stupidity. If innate dullness was to be the "official" scientific rationale for excluding immigrants from eastern and southern Europe, how could the Jews be kept out?

The most attractive possibility lay in claiming that the old catalog had been too generous and that, contrary to its popular stereotype, Jews were stupid after all. Several "scientific" studies conducted between 1910 and 1930, the heyday of the great immigration debate, reached this devoutly desired conclusion. As examples of distorting facts to match expectations or of blindness to obvious alternatives, they are without parallel. This essay is the story of two famous studies, from different nations and with different impact.

H. H. Goddard was the director of research at the Vineland Institute for Feebleminded Girls and Boys in New Jersey. He viewed himself as a taxonomist of mental deficiency. He concentrated upon "defectives of high grade" who posed special problems because their status just below the borderline of normality rendered their identification more difficult. He invented the term "moron" (from a Greek word for "foolish") to describe people in this category. He believed at the time, although he changed his mind in 1928, that most morons should be confined to institutions for life, kept happy with tasks apportioned to their ability, and above all, prevented from breeding.

Goddard's general method for identifying morons was simplicity itself. Once you had enough familiarity with the beast, you simply looked at one, asked a few questions, and drew your evident conclusions. If they were dead, you asked questions of the living who knew them. If they were dead, or even fictitious, you just looked. Goddard once attacked the poet Edwin Markham for suggesting that "The Man with the Hoe," inspired by Millet's famous painting of a peasant, "came to his condition as the result of social conditions which held him down and made him like the clods that he turned over." Couldn't Markham see that Millet's man was mentally deficient? "The painting is a perfect picture of an imbecile," Goddard remarked. Goddard thought he had a pretty good eye himself, but the main task of identifying morons must be given to women because nature had endowed the fair sex with superior intuition:

> After a person has had considerable experience in this work, he almost gets a sense of what a feeble-minded person is so that he can tell one afar off. The people who are best at this work, and who I believe should do this work, are women. Women seem to have closer observation than men.

In 1912, Goddard was invited by the U.S. Public Health Service to try his skill at identifying morons among arriving immigrants on Ellis Island. Perhaps they could be screened out and sent back, thus reducing the "menace of the feebleminded." But this time, Goddard brought a new method to supplement his identifications by sight—the Binet tests of intelligence, later to become (at the hands of Lewis M. Terman of Stanford University), the Stanford-Binet scale, or the conventional measure of IQ. Binet had just died in France and would never witness the distortion of his device for identifying children who needed special help in school into an instrument for labeling people with a permanent stamp of inferiority.

Goddard was so encouraged by the success of his preliminary trials that he raised some money and sent two of his women back to Ellis Island in 1913 for a more thorough study. In two and a half months, they tested four major groups: thirty-five Jews, twenty-two Hungarians, fifty Italians, and forty-five Russians. The Binet tests produced an astounding result: 83 percent of the Jews, 87 percent of the Russians, 80 percent of the Hungarians, and 79 percent of the Italians were feebleminded—that is, below mental age twelve (the upper limit of moronity by Goddard's definition). Goddard himself was a bit embarrassed by his own exaggerated success. Weren't his results too good to be true? Could people be made to believe that four-fifths of any nation were morons? Goddard played with the numbers a bit, and got his figures down to 40 or 50 percent, but he was still perturbed.

The Jewish sample attracted his greatest interest for two reasons. First, it might resolve the dilemma of the supposedly intelligent Jew and provide a rationale for keeping this undesirable group out. Second, Goddard felt that he could not be accused of bias for the Jewish sample. The other groups had been tested via interpreters, but he had a Yiddish-speaking psychologist for the Jews.

In retrospect, Goddard's conclusions were far more absurd than even he allowed himself to suspect in anxious moments. It became clear, a few years later, that Goddard had constructed a particularly harsh version of the Binet tests. His scores stood well below the rankings produced by all other editions. Fully half the people who scored in the low, but normal, range of the Stanford-Binet scale tested as morons on Goddard's scales.

But the greater absurdity arose from Goddard's extraordinary insensitivity to environmental effects, both long-term and immediate, upon test scores. In his view, the Binet tests measured innate intelligence by definition, since they required no reading or writing and made no explicit reference to particular aspects of specific cultures. Caught in this vicious circle of argument, Goddard became blind to the primary reality that surrounded his women on Ellis Island. The redoubtable Ms. Kite approaches a group of frightened men and women—mostly illiterate, few with any knowledge of English, all just off the boat after a grueling journey in steerage—plucks them from the line and asks them to name as many objects as they can, in their own language, within three minutes. Could their poor performance reflect fear, befuddlement, or physical weakness rather than stupidity? Goddard considered the possibility but rejected it:

> What shall we say of the fact that only 45 percent can give sixty words in three minutes, when normal children of 11 years sometimes give 200 words in that time! It is hard to find an explanation except lack of intelligence. . . . How could a person live even 15 years in any environment without learning hundreds of

> names of which he could certainly think of 60 in three minutes.

Could their failure to identify the date, or even the year, be attributed to anything other than moronity?

> Must we again conclude that the European peasant of the type that immigrates to America pays no attention to the passage of time? That the drudgery of life is so severe that he cares not whether it is January or July, whether it is 1912 or 1906? Is it possible that the person may be of considerable intelligence and yet, because of the peculiarity of his environment, not have acquired this ordinary bit of knowledge, even though the calendar is not in general use on the continent, or is somewhat complicated as in Russia? If so what an environment it must have been!

Goddard wrestled with the issue of this moronic flood. On the one hand, he could see some benefits:

> They do a great deal of work that no one else will do. . . . There is an immense amount of drudgery to be done, an immense amount of work for which we do not wish to pay enough to secure more intelligent workers. . . . May it be that possibly the moron has his place.

But he feared genetic deterioration even more and eventually rejoiced in the tightening of standards that his program had encouraged. In 1917, he reported with pleasure that deportations for mental deficiency had increased by 350 percent in 1913 and 570 percent in 1914 over the average for five preceding years. Morons could be identified at ports of entry and shipped back, but such an inefficient and expensive procedure could never be instituted as general policy. Would it not be better simply to restrict immigration from nations teeming with morons? Goddard suggested that his conclusions "furnish important considerations for future actions both scientific and social as well as legislative." Within ten years, restriction based upon national quotas had become a reality.

Meanwhile, in England, Karl Pearson had also decided to study the apparent anomaly of Jewish intelligence. Pearson's study was as ridiculous as Goddard's, but we cannot attribute its errors (as we might, being unreasonably charitable, in Goddard's case) to mathematical naïveté, for Pearson virtually invented the science of statistics. Pearson, the first Galton Professor of Eugenics at University College, London, founded the *Annals of Eugenics* in 1925. He chose to initiate the first issue with his study of Jewish immigration, apparently regarding it as a model of sober science and rational social planning. He stated his purpose forthrightly in the opening lines:

> The purport of this memoir is to discuss whether it is desirable in an already crowded country like Great Britain to permit indiscriminate immigration, or, if the conclusion be that it is not, on what grounds discrimination should be based.

If a group generally regarded as intellectually able could be ranked as inferior, then the basic argument for restriction would be greatly enhanced, for who would then defend the groups that everyone considered as stupid? Pearson, however, loudly decried any attempt to attribute motive or prior prejudice to his study. One can only recall Shakespeare's line, "The lady doth protest too much, methinks."

> There is only one solution to a problem of this kind, and it lies in the cold light of statistical inquiry. . . . We have no axes to grind, we have no governing body to propitiate by well advertised discoveries; we are paid by nobody to reach results of a given bias. We have no electors, no subscribers to encounter in the market place. We firmly believe that we have no political, no religious and no social prejudices. . . . We rejoice in

> numbers and figures for their own sake and, subject to human fallibility, collect our data—as all scientists must do—to find out the truth that is in them.

Pearson had invented a statistic so commonly used today that many people probably think it has been available since the dawn of mathematics—the correlation coefficient. This statistic measures the degree of relationship between two features of a set of objects: height versus weight or head circumference versus leg length in a group of humans, for example. Correlation coefficients can range as high as 1.0 (if taller people are invariably heavier to the same degree) or as low as 0.0 for no correlation (if an increase in height provides no information about weight—a taller person may weigh more, the same, or less, and no prediction can be made from the increase in height alone). Correlation coefficients can also be negative if increase in one variable leads to decrease in the other (if taller people generally weigh less, for example). Pearson's study of Jewish immigration involved the measurement of correlations between a large and motley array of physical and mental characters for children of Jewish immigrants living in London.

Pearson measured everything he imagined might be important in assessing "worthiness." He established four categories for cleanliness of hair: very clean and tidy, clean on the whole, dirty and untidy, and matted or verminous. He assessed both inner and outer clothing on a similar scale: clean, a little dirty, dirty, and filthy. He then computed correlation coefficients between all measures and was generally disappointed by the low values obtained. He could not understand, for example, why cleanliness of body and hair correlated only .2615 in boys and .2119 in girls, and mused:

> We should naturally have supposed that cleanliness of body and tidiness of hair would be products of maternal environment and so highly correlated. It is singular that they are not. There may be mothers who consider

> chiefly externals, and so press for tidiness of hair, but it is hard to imagine that those who emphasize cleanliness of body overlook cleanliness of hair.

Pearson concluded his study of physical measures by proclaiming Jewish children inferior to the native stock in height, weight, susceptibility to disease, nutrition, visual acuity, and cleanliness:

> Jewish alien children are not superior to the native Gentile. Indeed, taken all round we should not be exaggerating if we asserted that they were inferior in the great bulk of the categories dealt with.

The only possible justification for admitting them lay in a potentially superior intelligence to overbalance their physical shortcomings.

Pearson therefore studied intelligence by the same type of short and subjective scale that had characterized his measures of physical traits. For intelligence, he relied upon teachers' judgments rated from A to G. Computing the raw averages, he found that Jewish children were not superior to native Gentiles. Jewish boys ranked a bit higher, but the girls scored notably lower than their English classmates. Pearson concluded, with a striking analogy:

> Taken on the average, and regarding both sexes, this alien Jewish population is somewhat inferior physically and mentally to the native population. . . . We know and admit that some of the children of these alien Jews from the academic standpoint have done brilliantly; whether they have the staying power of the native race is another question. No breeder of cattle, however, would purchase an entire herd because he anticipated finding one or two fine specimens included in it; still less would he do it, if his byres and pastures were already full.

But Pearson realized that he was missing one crucial argument. He had already admitted that Jews lived in relative poverty. Suppose intelligence is more a product of environment than inborn worth? Might not the average scores of Jews reflect their disadvantaged lives? Would they not be superior after all if they lived as well as the native English? Pearson recognized that he had to demonstrate the innateness of intelligence to carry his argument for restricted immigration based on irremediable mediocrity.

He turned again to his correlation coefficients. If low intelligence correlated with measures of misery (disease, squalor, and low income, for example), then an environmental basis might be claimed. But if few or no correlations could be found, then intelligence is not affected by environment and must be innate. Pearson computed his correlation coefficients and, as with the physical measures, found very few high values. But this time he was pleased. The correlations produced little beyond the discovery that intelligent children sleep less and tend to breathe more through their nose! He concluded triumphantly:

> There does not exist in the present material any correlation of the slightest consequence between the intelligence of the child and its physique, its health, its parents' care or the economic and sanitary conditions of its home. . . . Intelligence as distinct from mere knowledge stands out as a congenital character. Let us admit finally that the mind of man is for the most part a congenital product, and the factors which determine it are racial and familial. . . . Our material provides no evidence that a lessening of the aliens' poverty, an improvement in their food, or an advance in their cleanliness will substantially alter their average grade of intelligence. . . . It is proper to judge the immigrant by what he is as he arrives, and reject or accept him then.

But conclusions based upon negative evidence are always suspect. Pearson's failure to record correlations between "intelligence" and environment might suggest the true absence of any relationship. But it might also simply mean that his measures were as lousy as the hair in his category 4.

Maybe a teacher's assessment doesn't record anything accurately, and its failure to correlate with measures of environment only demonstrates its inadequacy as an index of intelligence. After all, Pearson had already admitted that correlations between physical measures had been disappointingly small. He was too good a statistician to ignore this possibility. So he faced it and dismissed it with one of the worst arguments I have ever read.

Pearson gave three reasons for sticking to his claim that intelligence is innate. The first two are irrelevant: teachers' assessments correlate with Binet test scores, and high correlations between siblings and between parents and children also prove the innateness of intelligence. But Pearson had not given Binet tests to the Jewish children and had not measured their parents' intelligence in any way. These two claims referred to other studies and could not be transferred to the present case. Pearson appreciated this weakness and therefore advanced a third argument based upon internal evidence: intelligence (teachers' assessment) failed to correlate with environment but it did correlate with other "independent" measures of mental worth.

But what were these other independent measures? Believe it or not, Pearson chose "conscientiousness" (also based on teachers' assessments and scored as keen, medium, and dull), and rank in class. How else does a teacher assess "intelligence" if not (in large part) by conscientiousness and rank in class? Pearson's three measures—intelligence, conscientiousness, and rank in class—were redundant assessments of the same thing: the teachers' opinion of their students' worth. But we cannot tell whether these opinions record inborn capacities, environmental advantages, or teachers' prejudices. In any case, Pearson concluded with an appeal to bar all but the most intelligent of foreign Jews:

> For men with no special ability—above all for such men as religion, social habits, or language keep as a caste apart, there should be no place. They will not be ab-

> sorbed by, and at the same time strengthen the existing population; they will develop into a parasitic race.

Goddard's and Pearson's studies shared the property of internal contradictions and evident prejudice sufficient to dismiss all claims. But they differed in one important respect: social impact. Britain did not enact laws to restrict immigration by racial or national origin. But in America, Goddard and his colleagues won. Goddard's work on Ellis Island had already encouraged immigration officials to reject people for supposed moronity. Five years later, the army tested 1.75 million World War I recruits with a set of examinations that Goddard helped write and that were composed by a committee meeting at his Vineland Training School. The tabulations did not identify Jews per se but calculated "innate intelligence" by national averages. These absurd tests, which measured linguistic and cultural familiarity with American ways, (see my book, *The Mismeasure of Man,* W. W. Norton, 1981), ranked recent immigrants from southern and eastern Europe well below the English, Germans, and Scandinavians who had arrived long before. The average soldier of most southern and eastern European nations scored as a moron on the army tests. Since most Jewish immigrants arrived from eastern European nations, quotas based on country of origin eliminated Jews as surely as collegiate quotas based on geographical distribution once barred them from elite campuses.

When quotas were set for the Immigration Restriction Act of 1924, they were initially calculated at 2 percent of people from each nation present in America at the census of 1890, not at the most recent count of 1920. Since few southern and eastern Europeans had arrived by 1890, these quotas effectively reduced the influx of Slavs, Italians, and Jews to a trickle. Restriction was in the air and would have occurred anyway. But the peculiar character and intent of the 1924 quotas were largely a result of propaganda issued by Goddard and his eugenical colleagues.

What effect did the quotas have in retrospect? Allan

Chase, author of *The Legacy of Malthus,* the finest book on the history of scientific racism in America, has estimated that the quotas barred up to six million southern, central, and eastern Europeans between 1924 and the outbreak of World War II (assuming that immigration had continued at its pre-1924 rate). We know what happened to many who wanted to leave but had no place to go. The pathways to destruction are often indirect, but ideas can be agents as surely as guns and bombs.

23 | The Politics of Census

IN THE CONSTITUTION of the United States, the same passage that prescribes a census every ten years also includes the infamous statement that slaves shall be counted as three-fifths of a person. Ironically, and however different the setting and motives, black people are still undercounted in the American census because poor people in inner cities are systematically missed.

The census has always been controversial because it was established as a political device, not as an expensive frill to satisfy curiosity and feed academic mills. The constitutional passage that ordained the census begins: "Representatives and direct taxes shall be apportioned among the several states which may be included within this union, according to their respective numbers."

Political use of the census has often extended beyond the allocation of taxation and representation. The sixth census of 1840 engendered a heated controversy based upon the correct contention that certain black people had, for once, been falsely *over*counted. This curious tale illustrates the principle that copious numbers do not guarantee objectivity and that even the most careful and rigorous surveys are only as good as their methods and assumptions. (William Stanton tells the story in *The Leopard's Spots,* his excellent book on the history of scientific attitudes toward race in America during the first half of the nineteenth century. I have also read the original papers of the major protagonist, Edward Jarvis.)

The 1840 census was the first to include counts of the mentally ill and deficient, enumerated by race and by state. Dr. Edward Jarvis, then a young physician but later to become a national authority on medical statistics, rejoiced that the frustrations of inadequate data would soon be overcome. He wrote in 1844:

> The statistics of insanity are becoming more and more an object of interest to philanthropists, to political economists, and to men of science. But all investigations, conducted by individuals or by associations, have been partial, incomplete, and far from satisfactory. . . . They could not tell the numbers of any class or people, among whom they found a definite number of the insane. And therefore, as a ground of comparison of the prevalence of insanity in one country with that of another, or in one class or race of people with that in another, their reports did not answer their intended purpose.

Jarvis then praised the marshals of the 1840 census as apostles of the new, quantitative order:

> As these functionaries were ordered to inquire from house to house, and leave no dwelling—neither mansion nor cabin—neither tent nor ship unvisited and unexamined, it was reasonably supposed that there would be a complete and accurate account of the prevalence of insanity among 17 millions of people. A wider field than this had never been surveyed for this purpose in any part of the earth, since the world began. . . . Never had the philanthropist a better promise of truth hitherto undiscovered. . . . Many proceeded at once to analyze the tables, in order to show the proportion of lunacy in the various states, and among the two races, which constitute our population.

As scholars and ideologues of varying stripes scrutinized the tables, one apparent fact rose to obvious prominence in those troubled times. Among blacks, insanity struck free people in northern states far more often than it afflicted slaves in the South. In fact, one in 162 blacks was insane in free states, but only one in 1,558 in slave states. But freedom and the North posed no mental terror for whites, since their relative sanity did not differ in the North and South.

Moreover, insanity among blacks seemed to decrease in even gradation from the harsh North to the congenial South. One in 14 of Maine's black population was either insane or idiotic; in New Hampshire, one in 28; in Massachusetts, one in 43; in New Jersey, one in 279. In Delaware, however, the frequency of insanity among blacks suddenly nose-dived. As Stanton writes: "It appeared that Mason and Dixon had surveyed a line not only between Maryland and Pennsylvania but also—surely all unwitting—between Sanity and Bedlam."

In his first publication on the 1840 census, Jarvis drew the same conclusion that so many other whites would advance: slavery, if not the natural state of black people, must have a remarkably beneficent effect upon them. It must exert "a wonderful influence upon the development of moral faculties and the intellectual powers." A slave gains equanimity by "refusing many of the hopes and responsibilities which the free, self-thinking and self-acting enjoy and sustain," for bondage "saves him from some of the liabilities and dangers of active self-direction."

The basic "fact" of ten times more insanity in freedom than in slavery was widely bruited about in the contemporary press, often in lurid fashion. Stanton quotes a contributor to the *Southern Literary Messenger* (1843) who, concluding that blacks grow "more vicious in a state of freedom," painted a frightful picture of Virginia should it ever become a free state, with "all sympathy on the part of the master to the slave ended." He inquired:

> Where should we find penitentiaries for the thousands of felons? Where, lunatic asylums for the tens of thousands of maniacs? Would it be possible to live in a

country where maniacs and felons meet the traveler at every crossroad.

But Jarvis was troubled. The disparity between North and South made sense to him, but its extent was puzzling. Could slavery possibly make such an enormous difference? If the information had not been stamped with a governmental imprimatur, who could have believed it? Jarvis wrote:

> This was so improbable, so contrary to common experience, there was in it such a strong prima facie evidence of error, that nothing but a document, coming forth with all the authority of the national government, and "corrected in the department of state," could have gained for it the least credence among the inhabitants of the free states, where insanity was stated to abound so plentifully.

Jarvis therefore began to examine the tables and was shocked by what he discovered. Somehow, and in a fashion that could scarcely represent a set of random accidents, the number of insane blacks had been absurdly inflated in reported figures for northern states. Jarvis discovered that twenty-five towns in the twelve free states contained not a single black person of sound mind. The figure for "all blacks" had obviously been recopied or misplaced in the column for "insane blacks." But data for 135 additional towns (including thirty-nine in Ohio and twenty in New York) could not be explained so easily, for these towns actually reported a population of insane blacks greater than the total number of blacks, both sane and unhinged!

In a few cases, Jarvis was able to track down the source of error. Worcester, Massachusetts, for example, reported 133 insane in a total black population of 151. Jarvis inquired and discovered that these 133 people were white patients living in the state mental hospital located there. With this single correction, the first among many, black insanity in Massachusetts dropped from one in 43 to one in 129. Jarvis, demoralized and angry, began a decade of unsuccessful campaigning to win an official retraction or correction of the 1840 census. He began:

> Such a document as we have described, heavy with its errors and its misstatements, instead of being a messenger of truth to the world to enlighten its knowledge and guide its opinions, it is, in respect to human ailment, a bearer of falsehood to confuse and mislead.

This debate was destined for a more significant fate than persistent bickering in literary and scholarly journals. For Jarvis's disclosures caught the ear of a formidable man: John Quincy Adams, then near eighty, and capping a distinguished career as leader of antislavery forces in the House of Representatives. But Adams's opponent was equally formidable. At that time, the census fell under the jurisdiction of the Department of State, and its newly appointed secretary was none other than John C. Calhoun, the cleverest and most vigorous defender of slavery in America.

Calhoun, in one of his first acts in office, used the incorrect but official census figures in responding to the expressed hope of the British foreign secretary, Lord Aberdeen, that slavery would not be permitted in the new republic (soon to be state) of Texas. The census proved, Calhoun wrote to Aberdeen, that northern blacks had "invariably sunk into vice and pauperism, accompanied by the bodily and mental afflictions incident thereto," while states that had retained what Calhoun called, in genteel euphemism, "the ancient relation" between races, contained a black population that had "improved greatly in every respect—in number, comfort, intelligence, and morals."

Calhoun then proceeded to evade the official request from the House, passed on Adams's motion, that the secretary of state report on errors in the census and steps that would be taken to correct them. Adams then accosted Calhoun in his office and recorded the secretary's response in his diary:

> He writhed like a trodden rattlesnake on the exposure of his false report to the House that no material errors had been discovered in the printed Census of 1840, and finally said that there were so many errors they balanced one another, and led to the same conclusion as if they were all correct.

Jarvis, meanwhile, had enlisted the support of the Massachusetts Medical Society and the American Statistical Association. Armed with new data and support, Adams again persuaded the House to ask Calhoun for an official explanation. And again Calhoun wriggled out, finally delivering a report full of obfuscation and rhetoric, and still citing the 1840 figures on insanity as proof that freedom would be "a curse instead of a blessing" for black slaves. Jarvis lived until 1884 and assisted in the censuses of 1850, 1860, and 1870. But he never won official rectification of the errors he had uncovered in the 1840 census; the finagled, if not outrightly fraudulent, data on insanity among blacks continued to be cited as an argument for slavery as the Civil War approached.

There is a world of difference between the overcount of insane blacks in 1840 and the undercount of poor blacks (and other groups) in central cities in 1980. First, although the source of error in the 1840 census has never been determined, we may strongly suspect some systematic, perhaps conscious manipulation by those charged with putting the raw data in tabular form. I think we can be reasonably confident that, with automated procedures and more deliberate care, the systematic errors of the 1980 census are at least honest ones. Second, the politics of 1840 left few channels open to critics, and Calhoun's evasive stubbornness finally prevailed. Today, nearly every census is subjected to legal scrutiny and challenge.

Yet behind these legal struggles stands the fact that we still do not know how to count people accurately. Voluminous numbers and extensive tabulation do not guarantee objectivity. If you can't find people, you can't count them

—and the American census is, by law, an attempt at exhaustive counting, not a statistical operation based on sampling.

If it were possible (however expensive) to count everyone with confidence, then no valid complaint could be raised. But it is not, and the very attempt to do so engenders a systematic error that guarantees failure. For some people are much harder to find than others, either by their direct resistance to being counted (illegal aliens, for example) or by the complex of unfortunate circumstances that renders the poor more anonymous than other Americans.

Regions with a concentration of poor people will be systematically undercounted, and such regions are not spread across America at random. They are located in the heart of our major cities. A census that assesses population by direct counting will be a source of endless contention so long as federal money and representation in Congress reach cities as rewards for greater numbers.

Censusing has always been controversial, especially since its historical purpose has usually involved taxation or conscription. When David, inveigled by Satan himself, had the chutzpah to "number" Israel (I Chronicles, chapter 21), the Lord punished him by offering some unpleasant alternatives: three years of famine, three months of devastation by enemy armies, or three days of pestilence (all reducing the population, perhaps to countable levels). The legacy of each American census seems to be ten years of contention.

6 | Extinction

24 Phyletic Size Decrease in Hershey Bars

THE SOLACE OF MY YOUTH was a miserable concoction of something sweet and gooey, liberally studded with peanuts and surrounded by chocolate—real chocolate, at least. It was called "Whizz" and it cost a nickel. Emblazoned on the wrapper stood its proud motto in rhyme—"the best nickel candy there izz." Sometime after the war, candy bars went up to six cents for a time, and the motto changed without fanfare—"the best candy bar there izz." Little did I suspect that an evolutionary process, persistent in direction and constantly accelerating, had commenced.

I am a paleontologist—one of those oddballs who parlayed his childhood fascination for dinosaurs into a career. We search the history of life for repeated patterns, mostly without success. One generality that works more often than it fails is called "Cope's rule of phyletic size increase." For reasons yet poorly specified, body size tends to increase fairly steadily within evolutionary lineages. Some have cited general advantages of larger bodies—greater foraging range, higher reproductive output, greater intelligence associated with larger brains. Others claim that founders of long lineages tend to be small, and that increasing size is more a drift away from diminutive stature than a positive achievement of greater bulk.

The opposite phenomenon of gradual size decrease is surpassingly rare. There is a famous foram (a single-celled marine creature) that got smaller and smaller before disappearing entirely. An extinct, but once major group, the graptolites (floating, colonial marine organisms, perhaps related to vertebrates) began life with a large number of stipes (branches bearing a row of individuals). The number of stipes then declined progressively in several lineages, to eight, four, and two, until finally all surviving graptolites possessed but a single stipe. Then they disappeared. Did they, like the *Incredible Shrinking Man* simply decline to invisibility—for he, having decreased enough to make his final exit through the mesh of a screen in his movie début, must now be down to the size of a muon, but still, I suspect, hanging in there. Or did they snuff it entirely, like the legendary Foo-Bird who coursed in ever smaller circles until he flew up his own you-know-what and disappeared. What would a zero-striped graptolite look like? In any case, they are no longer part of our world.

The rarities of nature are often commonplaces of culture; and phyletic size decrease surrounds us in products of human manufacture. Remember the come-on, once emblazoned on the covers of comic books—"52 pages, all comics." And they only cost a dime. And remember when large meant large, rather than the smallest size in a sequence of detergent or cereal boxes going from large to gigantic to enormous.

Consider the Hershey Bar—a most worthy standard bearer for the general phenomenon of phyletic size decrease in manufactured goods. It is the unadvertised symbol of American quality. It shares with Band-Aids, Kleenex, Jell-o and the Fridge that rare distinction of attaching its brand name to the generic product. It has also been shrinking fast.

I have been monitoring informally, and with distress, this process for more than a decade. Obviously, others have followed it as well. The subject has become sufficiently sensitive that an official memo emanated in December 1978 from corporate headquarters at 19 East Chocolate Avenue —in Hershey, Pa. of course. Hershey chose the unmodified hang-out and spilled all the beans, to coin an appropriate metaphor. This three page document is titled "Remember the nickel bar?" (I do indeed, and ever so fondly, for I started to chomp them avidly in an age of youthful innocence, ever so long before I first heard of the nickel bag.) Hershey defends its shrinking bars and rising prices as a strictly average (or even slightly better than average) response to general inflation. I do not challenge this assertion since I use the bar as a synecdoche for general malaise—as an average, not an egregious, example.

I have constructed the accompanying graph from tabular data in the Hershey memo, including all information from mid-1965 to now. As a paleontologist used to interpreting evolutionary sequences, I spy two general phenomena: gradual phyletic size decrease within each price lineage, and occasional sudden mutation to larger size (and price) following previous decline to dangerous levels. I am utterly innocent of economics, the dismal science. For me, bulls and bears have four legs and are called *Bos taurus* and *Ursus arctos.* But I think I finally understand what an evolutionist would call the "adaptive significance" of inflation. Inflation is a necessary spin-off, or by-product, of a lineage's successful struggle for existence. For this radical explanation of

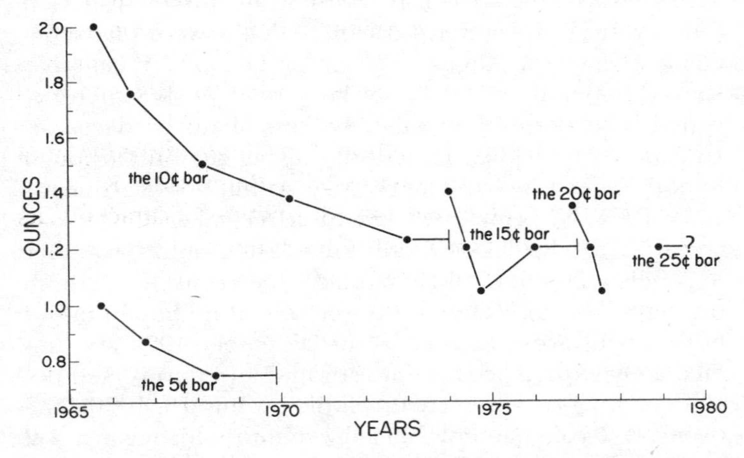

Hershey Bars bite the dust, a quantitative assessment. GRAPH BY L. MESZOLY.

inflation, you need grant me only one premise—that the manufactured products of culture, as fundamentally unnatural, tend to follow life's course in reverse. If organic lineages obey Cope's rule and increase in size, then manufactured lineages have an equally strong propensity for decreasing in size. Therefore, they either follow the fate of the Foo-Bird and we know them no longer, or they periodically restore themselves by sudden mutation to larger size —and, incidentally, fancier prices.

We may defend this thesis by extrapolating the tendencies of each price lineage on the graph. The nickel bar weighed an ounce in 1949. And it still weighed an ounce (following some temporary dips to ⅞ oz.) when our story began in September 1965. But it could delay its natural tendency no longer and decline began, to ⅞ oz. in September 1966 and finally to ¾ oz. in May 1968 until its discontinuation on November 24, 1969, a day that will live in infamy. But just as well, for if you extrapolate its average rate of decline (¼ ounce in thirty-two months), it would have become extinct naturally in May 1976. The dime bar followed a similar course, but beginning larger, it held on longer. It went steadily down from 2 oz. in August 1965 to 1.26 oz. in January 1973. It was officially discontinued on January 1, 1974, though I calculate that it would have become extinct on August 17, 1986. The fifteen-cent bar started hopefully at 1.4 oz. in January 1974, but then declined at an alarming rate far in excess of any predecessor. Unexpectedly, it then rallied, displaying the only (though minor) reverse toward larger size within a price lineage since 1965. Nonetheless, it died on December 31, 1976—and why not, for it could only have lasted until December 31, 1988, and who would have paid fifteen cents for a crumb during its dotage? The twenty-cent bar (I do hope I'm not boring you) arose at 1.35 oz. in December 1976 and immediately experienced the most rapid and unreversed decline of any price lineage. It will die on July 15, 1979. The twenty-five-cent bar, now but a few months old, began at 1.2 oz. in December 1978. *Ave atque vale.*

The graph shows another alarming trend. Each time the

Hershey Bar mutates to a new price lineage, it gets larger, but never as large as the founding member of the previous price lineage. The law of phyletic size decrease for manufactured goods must operate across related lineages as well as within them—thus ultimately frustrating the strategy of restoration by mutational jump. The ten-cent bar began at 2 oz. and was still holding firm when our story began in late 1965. The fifteen-cent bar arose at 1.4 oz., the twenty-cent bar at 1.35 oz., and the quarter bar at 1.2 oz. We can also extrapolate this rate of decrease across lineages to its final solution. We have seen a decrease of 0.8 oz. in three steps over thirteen years and four months. At this rate, the remaining four and a half steps will take another twenty years. And that ultimate wonder of wonders, the weightless bar, will be introduced in December 1998. It will cost forty-seven and a half cents.

The publicity people at Hershey's mentioned something about a ten-pound free sample. But I guess I've blown it. Still, I would remind everyone of Mark Twain's comment that there are "lies, damned lies and statistics." And I will say this for the good folks in Hershey, Pa. It's still the same damned good chocolate, what's left of it. A replacement of whole by broken almonds is the only compromise with quality I've noticed, while I shudder to think what the "creme" inside a Devil Dog is made of these days.

Still, I guess I've blown it. Too bad. A ten-pound bar titillates my wildest fancy. It would be as good as the 1949 Joe DiMaggio card that I never got (I don't think there was one in the series). And did I ever end up with a stack of pink bubble gum sheets for the effort. But that's another tale, to be told through false teeth at another time.

Postscript

I wrote this article (as anyone can tell from internal evidence) early in 1979. Since then, two interesting events have occurred. The first matched my predictions with uncanny accuracy. For the second, that specter of all science, the Great Exception (capital G, capital E), intervened and I have been temporarily foiled. And—as an avid Hershey bar chomper—am I ever glad for it.

The twenty-five cent bar did just about what I said it would. It started at 1.2 oz. in December 1978, where I left it, and then plummeted to 1.05 oz. in March 1980 before becoming extinct in March 1982. But Hershey then added a twist to necessity when it replaced its lamented two-bit bar with the inevitable thirty-cent concoction. Previously, all new introductions had begun (despite their fancier prices) at lower weights than the proud first item of the previous price lineage. (I based my extrapolation to the weightless bar on this pattern.) But, wonder of wonders and salaam to the Great Exception, the thirty-cent bar began at a whopping 1.45 oz., larger than anything we've seen since the ten-cent bar of my long-lost boyhood.

As cynical readers might expect, a tale lies behind this peculiar move. In the *Washington Post* for July 11, 1982 (and with thanks to Ellis Yochelson for sending the article), Randolph E. Bucklin explains all under the title: "Candy Wars: Price Tactic Fails Hershey."

It seems that the good folks at (not on) Mars, manufacturers of Three Musketeers, Snickers, and M & M's, and Hershey's chief competitor, had made the unprecedented move of increasing the size of their quarter bars without raising prices. After a while, they snuck the price up to thirty cents but kept the new size. Hershey tried to hold the line with its shrinking quarter bars. But thousands of mom and pop stores couldn't be bothered charging a quarter for some bars and thirty cents for others (and couldn't remember which were Hershey's and which Mars's anyway)—and therefore charged thirty cents for both Mars's large bars and Hershey's minuscule offerings. Hershey's sales plummeted; finally, they capitulated to Mars's tactics, raising prices to thirty cents and beefing up sizes to Mars's level and above predictions of the natural trend.

As a scientist trained in special pleading, I have a ready explanation for the Great Exception. General trends have an intrinsic character; they continue when external conditions retain their constancy. An unanticipated and unpredictable catastrophe, like the late Cretaceous asteroid of the next essay, or the sneaky sales tactic of Mars and Co., resets the system, and all bets are off. Still, the greater inevitability prevails. The thirty-cent bar will diminish and restitutions at higher prices will shrink as well. The weightless bar may come a few years later than I predicted (even a bit past the millennium)—but I still bet ya it'll cost about four bits.

25 | The Belt of an Asteroid

THE TEN PLAGUES of Moses are the archetypal disasters of western thought. I am therefore not surprised that popular explanations for major catastrophes in the history of life have tended to follow their scenarios in spirit. The most famous (although not the most profound) mass extinction occurred some sixty-five million years ago at the close of the Cretaceous period. All surviving dinosaurs died, as did their giant cousins of the air (pterosaurs) and seas (ichthyosaurs, plesiosaurs, mosasaurs). The oceanic plankton virtually disappeared with dramatic suddenness at a boundary that geologists call the plankton line. Several major groups of marine invertebrates perished, including all ammonites and the curious rudistid clams, which looked like corals and formed reefs.

Evidence for the cause of this great dying is so sparse that speculation receives free rein (and reign). The primal scenarios of Moses force themselves upon us. Theories of pandemics run wild recall the murrain that killed Pharaoh's cattle and the boils "breaking forth upon man, and upon beast." Poisoning of the oceans by copper washed in from the land or by a lens of fresh water spreading out from a fractured arctic lake reminds us of the Nile turned to blood —"and the fish that were in the river died; and the river stank. . . ." Dramatic change of climate conjures up the great hailstorm that fell "upon man, and upon beast, and upon every herb of the field. . . ." Voracious predators and parasitic pestilences have their counterpart in the successive deluges of frogs, lice, flies, and locusts that Pharaoh endured. Even the slaying of the children recalls the common (if rather silly) speculation about primitive mammals happily munching on dinosaur eggs. The only Mosaic plague that has not been well represented among the catalog of late Cretaceous disasters is the great darkness that blanketed Egypt for three days—"even darkness which might be felt."

I am happy to report that this serious omission has now been rectified. I am even more delighted to report that this latest entry has a basis in evidence of an entirely new sort. It has made legitimate, for the first time, a large class of explanations heretofore characterized by perfect plausibility in theory combined with utter lack of confirming evidence—extraterrestrial events. Would you believe an asteroid so big that it hit the earth and threw up a dust cloud thick enough to block photosynthesis entirely for a decade? Pharaoh almost tossed in the towel after a mere three days. But let me bypass the asteroid for a few paragraphs to discuss some ground rules and principles about mass extinction.

The major theories of mass extinction can be divided into two groups according to their stance on each of two issues: *source* (within or outside the earth) and *rate* (truly sudden or only relatively rapid). The earth itself is a source for some proposed causes, either by such physical mechanisms as changing climates engendered by shifting continents, or such biological factors as disease, competition, and collapse of food chains. Extraterrestrial hypotheses have ranged from variation in solar output, to cosmic radiation from nearby supernovae, to impacts of various bodies. Speaking of rate, some theories posit not merely a relatively rapid blip in the vastness of time but true cataclysms, disasters on the short scale of a human life—impacts of extraterrestrial bodies, for example. Other theories invoke processes that would be too slow to note during a human lifetime, but do their work in thousands or even millions of years, against a backdrop of billions. Most of these noncatastrophic theories implicate changes of climate, including drops in sea level and the growth of glaciers.

Geologists, like all folks, have their prejudices. They prefer causes emanating from their own domain, the earth. Since Lyell's day they have been trained to view major change as the accumulation of small inputs based on processes that can be observed in the relatively calm geologic present. These preferences have combined to give cataclysmic extraterrestrial theories a poor shake. Yet I think that few geologists would deem it inherently impossible, or even unlikely, that the earth might have suffered grievous cosmic insults at infrequent intervals during its vast history.

But another reason, better than traditional prejudice, governs the low esteem of extraterrestrial catastrophes. Geologists have not known, even in principle, any way to obtain direct evidence for them. What direct sign would a supernova or pronounced variation in solar intensity impose upon the earth? Indeed, the traditional argument for zapping by cosmic rays from supernovae relies upon total lack of evidence—the fact of mass extinction accompanied by no recognized geologic agent that might have caused it! Thus, many geologists, including myself, have long found themselves in the uncomfortable position of viewing extraterrestrial catastrophes as inherently plausible but rooting strongly against them. For what good is a theory, even a correct theory, that can generate no confirming evidence? The asteroidal theory has changed all that.

The facts of the Cretaceous extinction exert constraints upon the types of theories we may propose to explain them. We know, for example, that the extinctions occurred throughout the world and in all major environments—land, air, and sea. This fact alone virtually invalidates the entire panoply of popular theories that would attribute the extinction of dinosaurs to a cause related only to their supposed lumbering inefficiency—mammals eating their eggs, flowering plants pumping too much oxygen into the atmosphere, hyperpituitarism arising from large size and leading to sterility. Any harebrained idea can win notoriety in a context of such public fascination. Someone once proposed in all seriousness that male dinosaurs simply became too heavy to mount their partners for sexual intercourse, although I could never figure why little *Velociraptor* became extinct along with its giant cousins (not to mention what the giant brontosaurs were doing during the 100 million years or so of their success). The primary fact of dinosaur extinction is its timing as part of a global mass dying. We need a general theory, not a set of facile speculations for single groups.

We also know that the Cretaceous extinction included some aspects of geologically sudden death and others of more lingering demise. For some groups, the final phase of the Cretaceous seems to have acted more as a *coup de grâce* than an exterminating angel. Dinosaurs and ammonites had been in decline for millions of years. The dinosaur fauna of the latest Cretaceous did not include one of everything around a waterhole (as the multicolor chart on my kid's wall suggests), but a sharply reduced assemblage consisting largely of *Tyrannosaurus, Triceratops,* and a few smaller creatures. We can also correlate these slower declines with some geological events often implicated in extinctions (but please remember that correlation need not imply cause). Sea level declined steadily throughout the late Cretaceous; the continuous seaway that had split North America in two, running from Alaska to the Gulf of Mexico, withdrew gradually in both directions. As sea level dropped and continents grew in height and extent, temperatures began a general decline that continued throughout the next seventy million years, culminating in our recent (and still uncompleted) cycle of glacial ages.

Falling sea level has accompanied nearly every mass extinction that the earth has suffered; this correlation is about the only aspect of mass extinction that evokes general agreement among geologists. Its negative effect upon biological diversity also makes sense—for falling seas drained the extensive but shallow continental shelves, thereby removing a large chunk of living space from the domain of shallow-water invertebrates, the dominant fauna of our fossil record. Harsher conditions then spread across the land as the increasingly erratic and generally colder weather of a more "continental" earth prevailed. I doubt that any dinosaur ever ate an ammonite (although the giant mosasaurs, overgrown varanid lizards, did), but the coordinated decline of both groups may be causally related to dropping sea levels.

Yet we cannot attribute the entire Cretaceous extinction to a gradually deteriorating climate. Something more dramatic must have happened, as the plankton line testifies. Perhaps this dramatic cause gained greatly in effect because more groups than usual were in decline and therefore susceptible to a *coup de grâce.* In this sense, any complete account of the Cretaceous extinction will probably include a complex combination of dramatic end superimposed upon general deterioration.

In any case, geologic evidence constrains us to look for a contributing cause that is worldwide in effect, able to exterminate groups in all major habitats, and geologically sudden for at least some of its results. Which brings me back to asteroids.

The asteroidal theory, like so many interesting hypotheses in science, had its root in a study with markedly different aims (you cannot actively look for the utterly unexpected). A team headed by Luis and Walter Alvarez at Berkeley, California, thought that they might use the amount of iridium in sediments as an indicator of their depositional rate. Iridium, a rare metallic element of the platinum group, is one to ten thousand times more abundant in asteroids and meteorites than in the earth's crust and upper mantle. (Since both the earth and meteorites congealed from the same source, we must assume that the earth as a whole contains as high a percentage of iridium as the meteorites. But the earth melted and differentiated, and such heavy, unreactive elements as iridium sank into the inaccessible central core. The smaller bodies that form meteorites and asteroids never differentiated and therefore maintain iridium in its primeval abundance.) Hence, most iridium in the earth's sediments comes from extraterrestrial sources. Working with the common assumption that meteorites and cosmic dust fall upon the earth in a fairly constant rain, the Alvarezes reasoned that sediments high in iridium must have formed slowly since relatively less earth-based debris had accumulated to dilute the cosmic influx.

But the Alvarezes were not prepared for the anomalously high concentrations of iridium that they found in two places: in the Umbrian Apennines of Italy and near Copenhagen. Iridium levels were 30 times higher than average in Italy and 160 times higher in Denmark. Moreover, an analysis of twenty-seven other elements in the Italian sample showed that none departed by more than a factor of 2 from "average behavior" in ordinary sediments. The anomaly involves iridium alone.

The Alvarezes wondered if they could apply the style of explanation they had originally favored: could sedimentation have been slow enough in these two places to yield such a high concentration of iridium from the normal cosmic rain alone? But they could find no evidence or even think of any reason for believing that these sediments had formed during a virtual shut-off of normal, depositional processes in the ocean. Instead, they were forced to reverse their perspective: sedimentation had been more or less normal; the iridium represented a genuine cosmic influx of unusual amount, not an undiluted gentle rain. The Alvarezes had another outstandingly good reason to favor such a reversal. Both samples came from thin clays deposited at the very top of the Cretaceous—coincident with the great extinction.

But what extraterrestrial source might have both produced the iridium and acted as a cause of the great extinction? The Alvarezes looked first to that venerable old standby of cosmic theories—the supernova that exploded near the earth and zapped our planet with so much cosmic radiation that many creatures mutated themselves out of existence. Yet, after flirting with the idea (partly in public) for a while, they dropped it—to my great delight, for it has never made any biological sense to me, despite its almost knee-jerk popularity in the "disaster literature."

Radiation increases the mutation rate and yields a population with more variation. But more variation per se leads neither to extinction by prevalence of monstrosities nor to unusually rapid rates of evolution, because evolutionary tempos seem to be controlled by a different force—natural selection. Ordinary populations possess enough variation (without external goosing) to permit evolutionary rates so rapid that they appear instantaneous in geologic perspective. Mutation rates so high that they kill animals directly (not through the passage of defective genes to offspring) require supernovae too close to our sun to be plausible, given the spacing of stars in our part of the galaxy.

The Alvarezes now cite three reasons for rejecting a supernova:

1. A supernova would also have produced a high concentration of an ion of plutonium (^{244}Pu), yet the Italian and Danish clays contain levels of this ion ten times below the predicted value for a supernova.
2. Iridium occurs as two common isotopes (^{191}Ir and ^{193}Ir). Since all the objects of our solar system had a common origin, the ratio of these two isotopes should be the same in meteorites and in the small amount of iridium indigenous to the earth's crust. Iridium formed in other stars might exhibit a different ratio. The ratio in the anomalous Italian and Danish samples matches that of the earth's indigenous iridium and probably came from our solar system.
3. In order to zap the earth with as much iridium as the Italian and Danish samples contain, the exploding star would have to be so close to our sun that the probability of such an event becomes too small to be believed.

Since the ratio of iridium ions led the Alvarezes to seek a source within our own solar system, they turned to objects that might hit the earth with reasonable probability and do sufficient damage. Most asteroids orbit the sun in the large space between Mars and Jupiter, but a few follow more erratic paths, and some, called Apollo objects, cross the earth's orbit in their wanderings. Since the asteroid Apollo was discovered in 1933, twenty-seven others that cross the earth's orbit have been sighted. Astronomers discover an average of four more each year, while two independent estimates yield about 700 for the probable number of

Apollo asteroids more than one kilometer in diameter. The Alvarezes conclude that occasional collisions between Apollo asteroids and the earth are inevitable.

In short, their scenario for the Cretaceous extinction calls upon the impact of an Apollo asteroid approximately ten kilometers in diameter. They calculate that such an object would have produced a crater more than 150 kilometers in diameter and injected so much dust into the atmosphere both from its own pulverization and by dismemberment of the earth around it that our entire planet became as dark as Egypt during Pharaoh's ordeal. Photosynthesis might have been completely suppressed for a decade or so, leading to immediate death of the photosynthetic plankton (with their short generation time measured in weeks) and a subsequent collapse of the oceanic food chain based upon them. Most species of large terrestrial plants might have survived through the dormancy of their seeds, but the parental plants themselves would have died and taken their dinosaur herbivores with them. The Alvarezes calculate that Apollo asteroids of this size could have hit the earth with sufficient frequency to cause the five major extinctions that have punctuated the history of life since the inception of an adequate fossil record some 600 million years ago.

This scenario also contains major problems. I am most disturbed by the special pleading required to make the pattern of extinctions come out right. I can buy the oceanic argument, but balk at the Alvarezes' attempt to explain why three terrestrial groups got through the great darkness relatively unscathed: plants, small vertebrates (mammals and birds), and nearshore vertebrates. In a classic case of eating and having their cake, they permit mammals to survive by eating seeds, yet call upon those very seeds to save the plant species that engendered them. Their explanation for nearshore vertebrates is grounded more in hope than in logic: survival "may be due to their ability to utilize food chains based on nutrients derived from decaying land plants carried by rivers to the shallow seas."

The Alvarezes are also suffering from too much of a good thing, for levels of iridium in the Danish sample are embarrassingly high. They calculate that average asteroidal iridium mixed with about 100 times its mass of earth material (the amount needed to conjure up the great dust cloud) should increase mean iridium levels in sediments by only about one-tenth the Danish amount. The Italian sample runs closer to expectations.

Other problems may be in store. One uncertain study argues, from the fine stratigraphy of magnetic reversals, that the plankton line is not coincident with dinosaur extinctions—a small effect for the more leisurely concept of geologic suddenness (which might span many thousand years), but potentially fatal to the Alvarezes' requirement for true simultaneity. I feel that the Alvarezes assume the proper attitude toward this report: they acknowledge it forthrightly, note its uncertainties, admit that its later confirmation would greatly weaken or destroy their hypothesis, and then predict—on the basis of their theory—that the report is both probably in error and greatly in need of more study.

But I care little whether the asteroidal scenario itself is correct. The remarkable aspect of the Alvarezes' work—the part that has produced buzzing excitement among my colleagues, rather than the ho-hum that generally accompanies yet another vain speculation—lies in their raw data on enhanced iridium at the very top of the Cretaceous. For the first time we now have the hope (indeed the expectation) that evidence for extraterrestrial causes of mass extinction might exist in the geologic record. The old paradox—that we must root against such plausible theories because we know no way to obtain evidence for them—has disappeared.

When I started my career as a paleontologist, I used to argue that mass extinctions might be rip-roaring fun to discuss but relatively unimportant for the ultimate disposition of life and its history. I was then caught up in some common prejudices about inherent, stately progress as a hallmark of life's history. The mass extinctions, I thought, might disrupt the process severely, setting the clock of progress far back. But time (to a geologist) is not a severe constraint; life would recover, moving on as before.

I can't think of any other idea I once held (after age five) that I now regard as so foolish (except the thought that the N.Y. Giants might catch the Brooklyn Dodgers in 1951, which they did—thanks again, Bobby Thomson). The history of life has some weak empirical tendencies, but it is not going anywhere intrinsically. Mass extinctions do not simply reset the clock; they create the pattern. They wipe out groups that might have prevailed for countless millenniums to come and create ecological opportunities for others that might never have gained a footing. And they do their damage largely without regard to perfection of adaptation (the most gorgeously designed photosynthetic plankter could not survive a great darkness, while some marginal competitor might squeak through and become a progenitor of the next dominant group).

Who knows? Without the great Cretaceous extinction, dinosaurs might have rallied and still dominate the earth (they had already lived far longer than the sixty-five million years since their demise). Mammals might still be a small group of ratlike creatures casting about for an occasional bit of protein in a triceratopsian egg.

Among the Cretaceous mammals that witnessed the great event, we know a single primate named *Purgatorius.* It may not have been the only member of our order, but there probably weren't many of us back then. Suppose that *Purgatorius* hadn't pulled through—and remember, it probably owed its continuation to luck or to adaptations not related to features we value in primates because we have capitalized upon them. Primates would not have reevolved. A giraffe on the plains might now be surveying the creation from on high and smugly regarding itself as the finest of all God's creatures. Our current existence is an extended function of enormous improbabilities. We may owe our evolution, in large part, to the great Cretaceous dying that cleared a path, yet spared our ancestors' lives to tread upon it. That asteroid may well have been the sine qua non of our existence.

Postscript

So much has been written and argued about the Alvarez hypothesis since this essay appeared in June 1980 that any attempt at complete (or even adequate) update would require a book in itself. I therefore decided to let the essay stand largely as it appeared; at least it may have some historical virtue as a comment upon the hypothesis as first presented. Since then, the primary gain in support for Alvarez has been the discovery of iridium spikes right at the Cretaceous–Tertiary boundaries in many other places beyond the original two spots mentioned in the essay. Some include very different environments (deep sea cores) and very different places (New Zealand, at the virtual antipode of the two original sites). All this augurs well for a hypothesis that requires a worldwide effect. The primary difficulty and source of debate has centered upon an issue prominently discussed in my essay (one that any professional paleontologist would—and did—raise). In their original article (see bibliography) Alvarez *et al.* strongly suggested that their asteroid might provide a complete explanation for the pattern of late Cretaceous extinctions. Yet the paleontological record indicates clearly (as critics have effectively established) that many groups meeting their end in the Cretaceous were in decline for millions of years before any late Cretaceous disaster could have struck the earth. In my view, this fact does not downplay the significance of a potential latest Cretaceous catastrophe; it does not even reduce such a catastrophe to the relatively insignificant status of *coup de grâce.* For, as I argue in the essay, these gradual declines—had no terminal catastrophe been superimposed upon them—would probably not have led to a major extinction that reset the pattern for life's subsequent history, including our own evolution. We need a complex and synergistic theory for the Cretaceous extinction.

In October 1981, virtually all leading experts on extraterrestrial impacts, iridium spikes, paleontological evidence, etc., met in Snowbird, Utah, to discuss both the Alvarez hypothesis and the general issue of the role of extraterrestrial catastrophes in geological history. The proceedings of this conference will be published by the Geological Society of America in late 1982 or early 1983. This volume will be an important source for any reader who wishes to pursue this fascinating subject further.

26 | Chance Riches

IN LITERATURE, the idea of randomness is often coupled with chaos, lawlessness, or disorder to conjure up a vision of ultimate terror. In the *Essay on Man,* Alexander Pope evokes the image of an entire universe falling apart if its lawful order should ever be broken.

> Let earth unbalanc'd from her orbit fly,
> Planets and suns run lawless thro' the sky;
> Let ruling angels from their spheres be hurl'd,
> Being on being wreck'd, and world on world;
> Heav'n's whole foundations to their center nod,
> And nature tremble to the throne of God.

Pope's facile and hopeful solution simply banished the very idea of chance by a decree of faith:

> All nature is but art unknown to thee;
> All chance, direction which thou canst not see;
> All discord, harmony not understood;
> All partial evil, universal good.

Chance also got Darwin into trouble more than 100 years later when he granted it an important place in his evolutionary theory. Many critics, impelled by a knee-jerk negativism toward randomness, made no attempt to understand the strictly limited role that Darwin had awarded it. They argued: Darwin must be wrong; nature is so harmonious, animals are so well designed. This order cannot be the work of chance.

Darwin would not have disagreed. He constructed his theory in two parts, letting chance prevail in the first but strictly excluding it from the second for the conventional reason that chance could not, in his opinion, yield the order so prevalent in our world. In Darwin's theory, populations must first develop a large amount of heritable variation to provide raw material for the later, directing action of natural selection. Darwin viewed this pool of variation as random with respect to the direction of adaptive change—that is, if a species would be better off at smaller size, variation tends to arise with equal frequency at sizes both larger and smaller than the current average. The raw material for evolutionary change—and the raw material only—arises by a process of random mutation.

Natural selection then enters for the second part, and it acts as a conventional, deterministic, directing force. Natural selection fashions the raw material of variation by preserving and fostering individuals that vary in adaptive directions. In our hypothetical population, small organisms will, on average, rear more successful offspring. Slowly, but inexorably, the average size of individuals in the population will decline.

Darwin expressed himself with admirable clarity, yet critics have been misunderstanding this fundamental point for more than a century, from the Rev. Adam Sedgwick (Darwin's geological mentor, not a dogmatic and antiscientific theologian) to Arthur Koestler. The litany is ever the same: Darwin must be wrong; order cannot arise from chance. Again, Darwin never said that it could. Chance only produces raw material; natural selection directs evolutionary change.

Evolutionary theory is now stirring from the strict Darwinism that has prevailed during the past thirty years or so. While critics have not seriously challenged Darwin's mechanism on its primary turf of explaining adaptation, they have rallied around a claim for pluralism. Must all evolutionary change be viewed as adaptation and ascribed to natural selection? Randomness has become a central focus for critics because Darwin's strict dichotomy seems to be breaking. Randomness may not act only in generating variation; it may be an important agent of evolutionary change as well. The specter of chance is now truly intruding where Darwin's critics had falsely detected it before. Given both the surpassingly poor reputation of randomness in general, and the specific Darwinian tradition of limiting its role to the production of raw material only, this development in evolutionary theory is both exciting and, to many, distressing.

I should enter some disclaimers, if only to reduce disturbance. Randomness is making its bid as an agent of evolutionary change, but it is not threatening natural selection in the realm of adaptation. The beauty and aerodynamic efficiency of a bird's wing, the grace and good design of a fish's fins are not lucky accidents. Also, I use the specific meaning that "random" has long maintained in evolutionary theory: to describe changes that arise with no determined orientation. I do not use it as a general metaphor for chaos, disorder, or incomprehensibility (more on this later).

Evolution operates at three major levels: populations change as certain genes become more or less common because individuals carrying them have more or less success in rearing offspring; new species arise by the splitting of descendant populations from their ancestors; and evolutionary trends occur because some species are more successful than others in branching and persisting. Randomness is challenging the determinism of natural selection as a cause of evolutionary change at all three levels.

The genetic structure of populations. When natural selection operates in its usual way, genetic variation is reduced: the fit arise, in part, by the elimination of the unfit. The total amount of genetic variation in a population should represent a balance between the addition of new variation by mutation and the removal of unfit variants by natural selection. Since we have stores of data on rates of mutation and selection, we may predict the upper limit of variation that a population can maintain *if* selection acts upon all genes.

Techniques for measuring the amount of genetic variation in natural populations have been available only for the past fifteen years. Their first and primary result came as a surprise to many geneticists: most populations maintain too much variation to support the usual claim that all genes are scrutinized by natural selection.

To be sure, natural selection does not always eliminate. In some cases, it may act to increase or maintain genetic variation. The most common mode of maintenance is called "heterozygote advantage." Suppose that a gene exists in two forms, a dominant *A* and a recessive *a*. In sexual species, each individual has two copies of the gene, one from each parent. Now suppose that the mixture, or so-called heterozygote *Aa*, has a selective advantage over either pure form, the double dominant *AA*, or the double recessive *aa*. In this case, selection will preserve both *A* and *a* by favoring the heterozygote *Aa* individuals.

But even these modes of preservation have their limits, and many geneticists feel that populations still maintain too much variation for selective control. If they are right (and the issue remains under intense debate), then we must face the possibility that many genes remain in populations because selection cannot "see" them, and therefore cannot either mark them for elimination or remove other variants by favoring them. In other words, many genes may be *neutral.* They may be invisible to natural selection and their increase or decrease may be a result of chance alone.

Since "change of gene frequencies in populations" is the "official" definition of evolution, randomness has transgressed Darwin's border and asserted itself as an agent of evolutionary *change.* (This process of random increase or decrease of frequency is called "genetic drift." Contemporary Darwinism has always recognized drift, but has proclaimed it an infrequent and unimportant process, mostly confined to tiny populations with little chance of evolutionary persistence. The newer theory of neutralism suggests that many, if not most, genes in large populations owe their frequency primarily to random factors.)

The origin of species. Species are defined as populations that are reproductively isolated from all others. Placed in contact with other populations, a true species will maintain itself as a separate evolutionary entity and will not amalgamate with another population by hybridization. The key question for the origin of new species then becomes: how does reproductive isolation evolve?

In the traditional view, an ancestral population is split by a geographical barrier (continents may drift apart, mountain ranges rise, or rivers alter their course, for example). The two descendant populations then evolve by natural selection to fit the local environments of their separated places. In due time the populations become so different that they will no longer be able to interbreed, should they reestablish contact. Reproductive isolation is a by-product of adaptive evolution by natural selection.

During the past decade, the predominance of this mode has been challenged by a variety of new proposals advocating an interesting twist or reversal of perspective. They all argue that reproductive isolation can arise rapidly as a result of historical accidents with no selective significance at all. In this case, reproductive isolation comes first. By establishing new and discrete units, it provides an opportunity for selection to work. The ultimate success of such a species may depend upon the later development of selected traits, but the act of speciation itself may be a random event.

Consider, for example, the process called chromosomal speciation. Taxonomists have discovered that many groups of closely related species do not differ much in form, behavior, or even general genetic composition. But they do display outstanding differences in the number and form of chromosomes, and these differences produce the reproductive isolation that maintains them as distinct species. An obvious, and old, hypothesis suggests that each new species arises when a major chromosomal change occurs by accident and manages to establish a new population.

But this hypothesis suffers an obvious difficulty: the major change arises in a single individual. With whom shall it breed? The hybrid offspring of this mutant and a normal member of the population will almost surely be at a strong selective disadvantage. The mutation will therefore be swiftly eliminated.

Recent work on the structure of populations indicates that some reasonably common forms of social organization might facilitate the origin of new species by rapid and accidental chromosomal change. If populations are "panmictic," that is, if each female has an equal chance of breeding with each available male, then the chromosomal mutant cannot spread. But suppose that populations are subdivided into small groups of kin that breed exclusively with each other for several generations. Suppose also that these kin groups are harems, with a dominant male maintaining several females, and with brother–sister mating among the offspring.

If a chromosomal mutation arises in the dominant male, all his children will carry both the mutant (from their single father) and the normal form (from their various mothers). They will probably be at a strong selective disadvantage relative to normal offspring in other harems. But this may not matter, especially if they have limited contact with other groups. The offspring may die at a higher rate, but we only require that a few survive to mate with their similarly afflicted siblings.

One-fourth of these second generation offspring will be "pure" and contain two copies of the chromosomal mutation. If they can recognize and mate preferentially with each other in the next generation, all offspring will be pure mutants and members of a new species already (if they continue to mate with each other and if any hybrid offspring with normal forms suffers a strong selective penalty). The new chromosome is selectively neutral; it provides neither advantages nor disadvantages in itself. By establishing itself rapidly and accidentally in a small group, it permits the origin of a new species, again by chance. The new species may require substantial adaptive retooling for its subsequent survival, but this is a different, and later, matter.

My colleague Guy Bush, of the University of Texas, tells

me that horses provide, circumstantially at least, a strong case for chromosomal speciation. They all maintain the harem structure of kin breeding. Their seven living species (two horses, two asses, and three zebras) all look and act pretty much alike, despite some outstanding differences in external color and pattern. But their chromosome numbers differ greatly and surprisingly, from thirty-two in one of the zebras to sixty-six in that paradigm of the unpronounceable, Przewalski's wild horse.

Major patterns of rise and fall in the history of life. Many readers might be willing to accept chance at the lower levels. A bit of slop for some "invisible" genes within populations, an accidental species here and there. But surely conventional reasons rule the grand ebb and flow of major groups in the history of life. Trilobites must not be as good as the "advanced" arthropods (shrimps and their allies) that now populate our seas. The hordes of brachiopods from times past must have been pushed out by other creatures, clams in particular, that look like them but work better. Dinosaurs must have been bad at something that mammals can do. At this level, natural selection must reign and life must improve.

The facts of mass extinction proclaim the fallacy of this argument. If groups slowly replaced other groups, some gaining in species over millions of years while others lost just as steadily, a scenario of selective control might seem irresistible. But most groups disappeared during the episodes of mass extinction that have punctuated the history of life. This fact is not news. It has been acknowledged for decades but explained away with an assumption that differential mortality must have a selective basis. The groups that roared (or even squeaked) through a holocaust must have survived for a reason. They were the tough guys, the good competitors. But some recent data on the extent of mass extinctions must call this comforting explanation into question.

The granddaddy of all extinctions occurred about 225 million years ago at the end of the Permian period. (We don't know why, although the coalescence of all continents at about the same time must have set the basic stage.) By wiping out many groups, permanently debilitating others, and allowing some to pass through relatively unscathed, this great dying set the fundamental pattern of life's diversity ever since. But how profound was the Permian extinction? An old and familiar figure states that half the families of marine organisms (52.0 percent to be exact) perished at that time. But families are taxonomic abstractions. What does 52 percent of families mean for species, nature's real units? (Inconsistent taxonomic practices and an inadequate fossil record prevent the counting of species directly. Families, as bigger units, are harder to miss.)

We can be sure that a removal of half the families requires the death of a much greater percentage of species. A family is not gone until all its species die, and many families contain tens or hundreds of species. The extinction of most individual species does not wipe out a family, just as, for example, the random excision of a single entry in a telephone directory rarely removes the name entirely—you'd have to bump off a lot of Smiths. How many species must die before 50 percent of families are gone?

David M. Raup of Chicago's Field Museum has recently considered this question (see bibliography). The problem has no easy solution. If all families contained about the same number of species, then a simple formula would suffice. But variation is enormous. Many families contain only a single species. In this case, removal of the species also wipes out the family. Phone books contain their Zzyzzymanskis as well as their Wongs. Other families contain more than 100 species. We must know the empirical distribution of species per family before we can make a proper estimate. And we cannot construct an empirical distribution for Permian families because we cannot count the species directly.

Raup therefore made tabulations for a group that we do know well, the echinoids, or sea urchins. Echinoids include 894 species distributed into 222 genera and 40 families. How many species must, on average, be removed at random in order to eliminate 52 percent of the families? Raup considered this question both empirically and theoretically and came up with the astounding figure of 96 percent. If the rest of life maintains a distribution of species within families similar to the echinoids—and we have no evidence for major differences in this pattern—then the Permian debacle might have wiped out all but 4 percent of species.

Since estimates of living species in the late Permian range from 45,000 to 240,000, a removal of 96 percent would leave but 1,800 to 9,600 species as guardians of life's continuity. Moreover, as Raup argues, we have no strong evidence, despite intensive and specific search, of selectivity in the Permian extinctions. The debacle did not seem to favor any particular kind of animal—bigger creatures, inhabitants of shallow water, more complex forms, for example.

I am not entirely persuaded by the 96 percent figure. Echinoids may not be a good model for all of life. More important, Raup is assuming that the 52 percent figure is not artificially inflated by biases in the fossil record. We know, for example, that late Permian marine sediments are relatively rare and we may be missing some successful families simply because so few late Permian fossils have been preserved. Nonetheless, even the most conservative figures indicate a removal of 80 to 85 percent of all species.

I think that we must therefore face an unpleasant fact. If anywhere near 96 percent of all species died, leaving as few as two thousand forms to propagate all subsequent life, then some groups probably died, and others survived, for no particular reason at all. Organisms can muster few defenses against a catastrophe of such magnitude, and survivors may simply be among the lucky 4 percent. Since the Permian extinction set the basic pattern of life's subsequent diversity (no new phyla and few classes have originated since then), our current panoply of major designs may not represent a set of best adaptations but a group of fortunate survivors.

If we must admit randomness as an important agent of evolutionary *change* at all levels, what shall we make of it? Shall we surrender to despair and proclaim the history of life both chaotic and unknowable? Such a solution might embody Pope's equation of chance with disorder, but it would represent a great misunderstanding of what randomness means, for two reasons.

First, chance may well describe a sequence of events without implying that each individual item has no cause. Take the classic random event, a coin flip or the throw of a die. I imagine that each flip has a determined outcome if we could (as we cannot) specify all the multifarious factors that enter into it—height above ground, force of the flip, initial side up, angle of first contact with the ground, for example. But the factors are too numerous and not under our control; an equal chance for each possible outcome is the best prediction we can make in the long run.

Perhaps the Permian extinction worked like a set of dice with few winning outcomes. Each species became extinct for a conventional local reason—this pond dried up, that estuary became too salty or suffered an invasion of a particularly efficient predator. But the reasons are so numerous and so beyond our ken that an equal chance for the removal of each species represents our best prediction for the overall outcome.

Only in this way could several inventors of probability theory stomach their own creation, for they were conventional believers in deterministic causation and convinced theists unwilling to debar purpose from the world. Charles Bell, author of a famous work on the human hand as a reflection of God's wisdom through its intricate design, wrote in 1833:

> We say, in common parlance, that the dice being shaken together, it is a matter of chance what faces they will turn up; but if we could accurately observe their position in the box before the shaking, the direction of the force applied, its quantity, the number of turns of the box, and the curve in which the motion was made, the manner of stopping the motion and the line in which the dice were thrown out, the faces turned up would be a matter of certain prediction.

This explanation may be comforting (and true), but I think we must face a second possibility. Perhaps randomness is not merely an adequate description for complex causes that we cannot specify. Perhaps the world really works this way, and many events are uncaused in any conventional sense of the word. Perhaps our gut feeling that it cannot be so reflects only our hopes and prejudices, our desperate striving to make sense of a complex and confusing world, and not the ways of nature.

What solace, then, can we have, if solace we need. An answer, I believe, lies in rejecting another traditional belief, Pope's false equation of chance with a host of fearful cognates: disorder, chaos, lawlessness, unpredictability, and destruction. For contrary to popular belief, random means none of these. It may not, as in Bell's dice, imply a lack of causation. And even if it does in many cases, a random process need not engender unpredictable disorder. Random processes can yield highly complex order. We have an elaborate theory for predicting the results of coin flipping, the archetypical random process. Suppose we flip six coins at once, over and over again. We can predict how many times the most common result of three heads and three tails will occur, and how many times the rare event of all heads or all tails will arise (one in sixty-four for each or one in thirty-two for either).

Admittedly, this is predictability of a different order. It works only for repeated trials in the long run. We can assign a probability to each individual flip, but we cannot determine a specific outcome. Nonetheless, the end result is ordered and predictable. Does not this kind of randomness offer sufficient comfort against the threat of chaos? Does it not even make for a more intriguing world? After all, it is chance, in this sense, that gives our own lives, and the course of human history, so much richness and interest. Call it by its older names of fortune or free will, if you like. Shall we deny a similar richness to the rest of nature?

27 O Grave, Where Is Thy Victory?

BILL LEE, certainly the most colorful if not the most skillful of baseball pitchers, once argued that his declining effectiveness on the mound could be traced to the "designated hitter" rule. (This rule, for you nonadepts, allows a manager in the American League to designate a permanent pinch hitter for any player in the regular lineup. Since most pitchers are hopeless at hitting, the designated hitter almost always substitutes for the pitcher, and pitchers, therefore, no longer come to bat.) "Every species that's become extinct," Lee proclaimed, "has done so because of overspecialization."

In this statement, baseball's self-styled philosopher repeats what may be the most common misconception about the history of life—that extinction is the ultimate sign of failure. No stigma seems to be greater than irrevocable disappearance. Dinosaurs dominated the land for 100 million years, yet a species that measures its own life in but tens of thousands has branded dinosaurs as a symbol of failure. Two years ago, for example, the good folks at Audi claimed (in unsubtle comparison with their large competitors) that *Brontosaurus* was "arguably the worst designed creature of all time." "Evolution," they continued, "has a sure way of correcting faulty design"—extinction, of course. Paleontologists rose in protest and Audi backed down. "You can be sure we will treat the *Brontosaurus* with more respect in the future," they promised.

I believe that this equation of disappearance with incompetence reflects an outmoded approach based on the false metaphor of progress and an overly grim view of natural selection as a persistent and endless life-or-death struggle among competitors—the military version of such Darwinian phrases as "survival of the fittest" and "the struggle for life." If life moves ever upward and onward by ruthless battle and elimination of losers, extinction must be the ultimate sign of inadequacy. But life is not a tale of progress; it is, rather, a story of intricate branching and wandering, with momentary survivors adapting to changing local environments, not approaching cosmic or engineering perfection. And success in natural selection is less the result of murder and mayhem than of producing more surviving offspring.

The equation of extinction with inadequacy makes no sense in the long view of paleontology. Extinction is the ultimate fate of all lineages, yet we surely cannot argue that all species are therefore badly designed or poorly adapted. Extinction is no shame. It is, in one sense, the enabling force of the biosphere. Since most species are extraordinarily resistant to major evolutionary change and since many habitats are fairly full of species, how could evolution proceed if extinction did not open space for novelty? Would I be writing, or you reading, if dinosaurs had survived and mammals remained, as they had for 100 million years, a minor group of small creatures living in ecological nooks and crannies that dinosaurs didn't penetrate?

If most extinctions were the direct result of competition with superior species, or even if most represented an inevitable failure to meet the challenges of minor environmental change (as Bill Lee charged), then a stigma might be attached to disappearance. But many, if not most, extinctions are reactions to environmental challenges so severe and unpredictable that we have no right to expect a successful response and, therefore, no reason to "blame" a species for its disappearance. A freshwater fish might dart and swim so elegantly that an engineer would proclaim its anatomy optimal. But if the lakes and rivers dry up, what defense does it have? Will blue whales be any less exquisite in design if rapacious humanity does the last one in? Some insurance policies offer no protection against cataclysms so momentous and unexpected that legal language calls them "acts of God." Species often die for reasons equally beyond control or calculation.

I may make these statements baldly, but you have no reason to accept them unless I can back them up with evidence and numbers. For the past decade, a group of researchers centered in Chicago (but admitting some outsiders like myself to marginal membership) has been working to quantify patterns of diversity in the history of life. These studies have provided our most extensive and consistent data on extinction. The findings support my central contention that extinction is no disgrace, but usually an inevitable result of circumstances beyond reasonable response. A pair of papers published in *Science* during March 1982 reach three conclusions central to this discussion.

1. *A quantification of mass extinction.*

We have known since the dawn of paleontology that extinctions are not spread evenly over time, but are concentrated in a few brief periods of markedly enhanced, often worldwide decimation—the so-called mass extinctions of the geological record. The boundaries of the geological time scale correspond with these epochs of extinction. (Each year, when my students groan at my request, or rather demand, that they memorize the geological time scale, I reply that those funny names—Cambrian, Ordovician, Silurian—are not capricious instruments of torture, but records of the outstanding events of life's history.)

D. M. Raup and J. J. Sepkoski have gathered and summarized the data of geological longevity for all forms of marine life. Their plot of extinction rates (families of organisms disappearing per million years) versus geological time offers few surprises in general outline, but provides our best and most consistent account of the *quantities* involved. Four brief periods of mass extinction stand well above the ordinary, or "background," rate of normal times, and a fifth nearly reaches the level of these major episodes. Two events mark the well-known era boundaries: the great Permian extinction, which may have extirpated more than ninety percent of shallow-water marine species some 225 million years ago (see essay 26), and the Cretaceous debacle, which wiped out remaining dinosaurs and a host of marine creatures some 65 million years ago (see essay 25). The three other events, although well enough known to paleontologists, are not emblazoned upon popular consciousness. Two occurred before the Permian (Ordovician and Devonian) and one between the Permian and Cretaceous (Triassic).

Raup and Sepkoski discovered that these brief mass extinctions are even more pronounced than previous data had suggested. The average background rate varies between 2.0 and 4.6 families per million years, while mass extinctions reach 19.3 families per million years. The authors conclude: "Our analysis shows that major mass extinctions are far more distinct from background extinction than has been indicated by previous analyses of other data sets."

Proposals for causes of these mass extinctions (see other essays in this section) range from continental coalescence and its sequelae (for the Permian) to asteroidal impact (for the Cretaceous)—causes that lie in the category of fluctuations beyond control or reasonable response and thus surrounding their victims with no aura of shame. Since these mass extinctions are even more massive than previously recognized, the scope of "blameless" extinction has been greatly widened.

2. *The supposedly classic case of extinction due to competitive inferiority cannot be supported.*

During most of Tertiary time (the "age of mammals") South America was an island continent—a sort of super Australia—with an indigenous fauna more than matching the marsupials down under for interest and peculiarity. The Australian region sports only one exclusive order of mammals, the egg-laying monotremes (echidna and duck-billed platypus). South America once housed several, with odd animals ranging from the rhinolike, but not rhino-related, toxodonts that Darwin discovered during his apprenticeship on the *Beagle,* to litopterns, which outhorsed horses by reducing toes from several to one and even losing the side splints (reduced vestiges of toes) that horses retain (see essay 14), to giant sloths and glyptodonts. Other oddities belonged to orders dwelling elsewhere but expressing a unique South American twist. All large mammalian carnivores, for example, were marsupials and included such outstanding creatures as the saber-toothed *Thylacosmilus.*

All these animals are gone today, victims of the greatest biological tragedy of the past five million years. For once, humans are absolved, and we must cite instead the rise of the Isthmus of Panama just a few million years ago. The isthmus connected South America with the more cosmopolitan fauna of northern continents. North American mammals came (wandering over the isthmus) and, in the traditional view, saw and conquered as well. What we usually regard as the modern "native" South American fauna—from llamas to alpacas to jaguars to tapirs to peccaries—are all relatively recent northern migrants.

The traditional view, with its odor of racist metaphor, pits a sleek, fine-tuned, and rigorously adapted northern fauna, tested in the crucible of harsh climates and relentless competition from previous waves of Asiatic and European migrants, against a lazy, stagnant, and unchallanged native South American fauna. What chance did the poor toxodonts and litopterns have? Superior northern forms came streaming over the isthmus and wiped them out. In return, only a few inferior South American forms managed to travel the other way and survive. We got porcupines, opossums, and the nine-banded armadillo. South America received an entire new regime.

If this tale be true, then perhaps battle is the law of life and extinction does connote defeat. But is it correct? Do the numbers support it? Did more northern forms go south than vice versa? Were extinction rates much higher for South American forms? L. G. Marshall and S. D. Webb joined Raup and Sepkoski in a second article that applies the same quantitative methods to the recent faunal history of South America. They conclude that several aspects of the traditional story are not true.

The interchange, first of all, was surprisingly symmetrical. Members of fourteen North American families now reside in South America, representing 40 percent of South American familial diversity. Twelve South American families now live in North America, forming 36 percent of North American families. At the finer scale of genera, reduction was also balanced on both sides of the isthmus. Native South American genera declined by 13 percent between pre- and post-isthmian faunas. Native North American genera declined by 11 percent during the same interval. Thus, about the same number of families moved successfully in both directions and about the same percentage of native forms became extinct on both sides following the initial interchange. Why then does the record carry its apparent message of North American victory and South American carnage?

I believe that three major reasons account for this impression, one social, one biological but largely spurious, and the third real and important. We must consider, first, the chauvinism of most Anglophones in the United States (whoops, I almost wrote Americans). Anything south of the Rio Grande is Spanish speaking and therefore culturally linked with South America. But a good part of North America lies between El Paso and Panama, and most South American migrants live there, not in the United States or Canada. After all, the equator runs through Quito (in a nation appropriately called Ecuador), and South America contains more tropical land than North America. Therefore, most migrating South American forms are tropical or subtropical in their climatic preferences, and their natural homes up north are Mexico or Central America. The paucity of South American migrants in my backyard (although I have seen an opossum) is no argument against their abundance or vigor.

Secondly, the taxonomic structure of South American forms dictated a greater effect upon overall diversity of design for an equal reduction in percentage of genera.

When the isthmus rose, many of South America's indigenous groups had already been reduced to such a low diversity that the removal of a genus or two extinguished the entire group. Few North American groups were tottering so near the brink. If one fauna has twenty groups with one genus each, and a second has two groups each with ten genera, then a removal of four genera from each fauna will wipe out four of the larger groups in one fauna and none in the other.

Finally, even though migrants moved with equal success in both directions and native forms declined in equal measure, North American migrants did fare "better" in one different and interesting way. When we count genera derived from migrants after they arrived in their new homes, we find an outstanding difference. In North America, genera originally from South America evolved very few new genera, while North American forms were remarkably prolific in South America. Twelve primary migrants from South America evolved but three secondary genera, while twenty-one North American migrants gave rise to forty-nine secondary genera in South America. Thus, North American forms radiated vigorously in South America and filled the continent with its modern fauna, while South American forms succeeded well enough in North America but did not radiate extensively.

Why this difference? The four authors suggest that a major phase in the rise of the Andes created a rain shadow over most of South America and led to the replacement of predominantly savanna-woodland habitats by drier forests and pampas and by deserts and semideserts in some areas. Perhaps North American forms radiated in a new habitat suited to their previous life styles, while South American forms continued to decline as their favored habitats shrank. Or perhaps the conventional explanation is true in part, and North American forms radiated because they are, in some unexplained way, competitively superior to South American natives—although most versions of "competitive superiority" will not explain higher rates of speciation but only success in battle (leading to longer duration of migrants coupled with higher rates of extinction among the vanquished, neither a component of this tale). In any case, the old story of "hail the conquering hero comes"—waves of differential migration and subsequent carnage—can no longer be maintained.

3. *A ray of comfort for the meliorists.*

The great eighteenth-century French naturalist George Buffon once expressed the fact of extinction in a fine literary image: "They must die out because time fights against them." We may now say, with equal literary license, that organisms may be fighting back. When Raup and Sepkoski compiled their data on the quantitative effect of mass extinctions, they made an interesting and unexpected discovery about the "background" level of normal times. They found that the background level has been slowly but steadily decreasing for more than a half billion years. In early Cambrian times, at the beginning of our adequate fossil record some 600 million years ago, the average rate stood at 4.6 families extinct per million years. Since then, the rate has declined steadily to about 2.0 families per million years today. If the Cambrian rate had been maintained, 710 more family extinctions would have occurred. It is intriguing—although I don't know what it means—that the total number of marine families has increased by almost the same number (680 families) since the Cambrian.

We must beware of reading too much into this arresting conclusion since, as Raup and Sepkoski remind us, biases of the fossil record must always be suspected as an artificial cause of such patterns. For example, probability of preservation increases for fossils in progressively younger rocks (greater geographic extent of sediments, less opportunity for destruction of fossils by subsequent heat and pressure). Perhaps older families seem to live for shorter times simply because records of their early or late appearance are not preserved. But if this pattern does reflect a biological reality, then it suggests that modern families are more resistant to extinction and that the total rise in life's diversity may be a result of this increasing general hardiness.

Still, heroic though the fight may be (in inappropriate metaphor), organisms cannot win. The rate of familial extinction may have been cut in half during life's recorded history, but no species is immortal, and all must ultimately perish. The perfection of immediate adaptation is no protection against massive fluctuations of environment that inevitably, in the course of millions of years, affect every corner of the globe. Since Darwinian processes can only improve local adaptations, and since species cannot reckon the future (with one interesting and imperfect exception), all will eventually perish, leaving as potential patrimony only the altered descendants that may branch off from them.

I was in York Cathedral this spring, where I found the essence of this theme expressed in charming doggerel on the seventeenth-century tomb of one William Gee. Sir William, it seems, lived such a blameless life that if God wished to confer immortality upon anyone, His candidate had surely come forth. But Sir William died, so God must not entertain this option:

> If universal learning, language, law
> Pure piety, religion's reverend awe,
> Fair friends, fair issue: if a virtuous wife,
> A quiet conscience, a contented life,
> The clergy's prayers or the poor man's tears
> Could have lent length to man's determined years.
> Sure as the fate, which for our fault we fear,
> Proud death had ne'er advanced this trophy here.
> In it behold thy doom, thy tomb provide.
> Sir William Gee had all these pleas, yet died.

(I have modernized spelling and punctuation, but not words or grammar. In line 8, read "would have" for the old subjunctive "had." Sir William's "issue" are his kids, so perhaps the family line survives.)

Inevitabilities should never be depressing. An old philosophical tradition, dating at least to Spinoza, proclaims that freedom is the recognition of necessity. If we respect intellect, true freedom must come from learning the ways of the world—what can be changed and what cannot—and by shaping a gutsy life accordingly. Besides, if species lived forever, we would have no science of paleontology, and I might have become a fireman after all.

7 | A Zebra Trilogy

28 What, If Anything, Is a Zebra?

EACH YEAR, professional scientists scan thousands of titles, read hundreds of abstracts, and study a few papers in depth. Since titles are the commonest, and usually the only, form of contact between writers and potential readers in the great glut of scientific literature, catchy items are appreciated and remembered, but unfortunately rare. Every scientist has his favorite title. Mine was coined by paleontologist Albert E. Wood in 1957: "What, If Anything, Is a Rabbit?"

Wood's question may have been wry, but his conclusion was ringing: rabbits and their relatives form a coherent, well-defined order of mammals, not particularly close to rodents in evolutionary descent. I was reminded of Wood's title recently when I read a serious challenge to the integrity of a personal favorite among mammals: the zebra. Now don't get too agitated. I am not trying to turn the world of received opinion upside down. Striped horses manifestly exist. But do they form a true evolutionary unit? With "Stripes Do Not a Zebra Make"—a quite respectable title in its own right—Debra K. Bennett has forced us to extend Wood's question to another group of mammals. What, if anything, is a zebra?

Since evolutionary descent is our criterion for biological ordering, we define groups of animals by their genealogy. We do not join together two distantly related groups because their members have independently evolved some similar features. Humans and bottle-nosed dolphins, for example, share the pinnacle of brain size among mammals. But we do not, for this reason, establish the taxonomic group Psychozoa to house both species—for dolphins are more closely related by descent with whales, and humans with apes. We follow the same principles in our own genealogies. A boy with Down's syndrome is still his parents' son and not, by reason of his affliction, more closely related to other Down's children, no matter how long the list of similar features.

The potential dilemma for zebras is simply stated: they exist as three species, all with black-and-white stripes to be sure, but differing notably in both numbers of stripes and their patterns. (A fourth species, the quagga, became extinct early in this century; it formed stripes only on its neck and forequarters.) These three species are all members of the genus *Equus,* as are true horses, asses, and donkeys. (In this essay, I use "horse" in the generic sense to specify all members of *Equus,* including asses and zebras. When I mean Old Dobbin or Man o' War, I will write "true horses.") The integrity of zebras then hinges on the answer to a single question: Do these three species form a single evolutionary unit? Do they share a common ancestor that gave rise to them alone and to no other species of horse? Or are some zebras more closely related by descent to true horses or to asses than they are to other zebras? If this second possibility is an actuality, as Bennett suggests, then horses with black-and-white stripes arose more than once within the genus *Equus,* and there is, in an important evolutionary sense, no such thing as a zebra.

But how can we tell, since no one witnessed the origin of zebra species (or at least australopithecines weren't taking notes at the time), and the fossil record is, in this case, too inadequate to identify events at so fine a scale. During the past twenty years, a set of procedures has been codified within the science of systematics for resolving issues of this kind. The method, called cladistics, is a formalization of procedures that good taxonomists followed intuitively but did not properly express in words, leading to endless quibbling and fuzziness of concepts. A clade is a branch on an evolutionary tree, and cladistics attempts to establish the pattern of branching for a set of related species.

Cladistics has generated a fearful jargon, and many of its leading exponents in America are among the most contentious scientists I have ever encountered. But behind the names and nastiness lies an important set of principles. Still the clear formulation of principles does not guarantee an unambiguous application in each case—as we shall see for our zebras.

I believe that we can get by with just two terms from the bounty offered by cladists. Two lineages sharing a common ancestor from which no other lineage has sprung form a *sister group.* My brother and I form a sister group (pardon the confusion of gender) because he is my only sib and neither of my parents had any other children.

Cladists attempt to construct hierarchies of sister groups in order to specify temporal order of branching in evolutionary history. For example: gorillas and chimpanzees form a sister group because no other primate species branched from their common ancestor. We may then take the chimp-gorilla sister group as a unit and ask which primate forms a sister group with it. The answer, according to most experts, is us. We now have a sister group with three species, each more closely related to its two partners than to any other species.

We may extend this process indefinitely to form a chart of

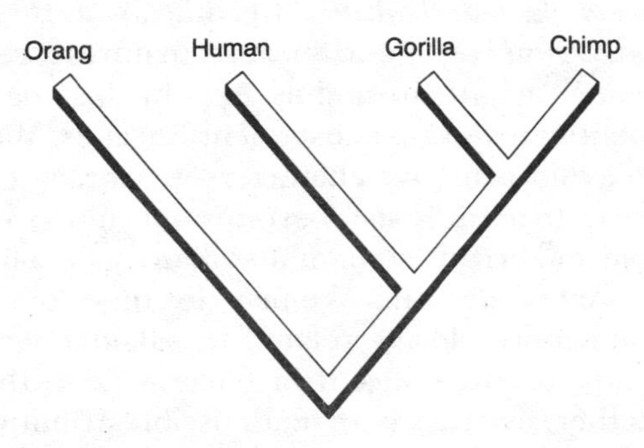

The cladistic pattern of great apes and humans. REPRINTED FROM NATURAL HISTORY. DRAWING BY JOE LE MONNIER.

branching relationships called a cladogram. But consider just one more step: What primate species is the sister group to the human-chimp-gorilla unit? Conventional wisdom cites the orangutan, and we add it to our cladogram.

This cladogram of "higher" primates contains an interesting implication: there is no such thing as an ape, at least as usually defined. Several species of primates may swing through trees, eat bananas in zoos, and form good prototypes for science fiction of various sorts. But orangs, chimps, and gorillas (the "apes" of our vernacular) are not a genealogical unit because orangs are cladistically more distant from chimps and gorillas than humans are—and we originally defined the term ape to contrast some lesser forms with our exalted state, not to include us!

The zebra problem can also be placed in this context. If the three species of zebras form a sister group (as humans, chimps, and gorillas do on our cladogram), then each is more closely related to its two partners than to any species of horse, and zebras form a true evolutionary unit. But if zebras are like "apes," and another species of horse lies within the cladogram of zebras (as humans lie within the cladogram of traditional apes), then striped horses may share some striking similarities meriting a common vernacular term (like zebra), but they are not a genealogical unit.

But how do we identify sister groups correctly? Cladists argue that we must search for—and here comes the second term—*shared derived characters* (technically called synapomorphies). Primitive characters are features present in a distant common ancestor; they may be lost or modified independently in several subsequent lineages. We must be careful to avoid primitive characters in searching for common features to identify sister groups, for they spell nothing but trouble and error. Humans and many salamanders have five toes; horses have one. We may not therefore state that humans are more closely related to salamanders than to horses, and that the concept of "mammal" is therefore a fiction. Rather, five toes is an inadmissible primitive character. The common ancestor of all terrestrial vertebrates had five toes. Salamanders and humans have retained the original number. Horses—and whales and cows and snakes and a host of other vertebrates—have lost some or all of their toes.

Derived characters, on the other hand, are features present *only* in members of an immediate lineage. They are unique and newly evolved. All mammals, for example, have hair; no other vertebrate does. Hair is a derived character for the class Mammalia because it evolved but once in the common ancestor of mammals and therefore identifies a true branch on the family tree of vertebrates. Shared derived characters are held in common by two or more lineages and may be used to specify sister groups. If we wish to identify the sister group among tunas, seals, and bobcats, we may use hair as a shared derived character to unite the two mammals and to eliminate the fish.

For zebras, the question then becomes: Are stripes a shared derived character of the three species? If so, the species form a sister group and zebras are a genealogical unit. If not, as Bennett argues, then zebras are a disparate group of horses with some confusing similarities.

The method of cladistics is both simple and sensible: establish sequences of sister groups by identifying shared derived characters. Unfortunately, conceptual elegance does not guarantee ease of application. The rub, in this case, lies in determining just what is or is not a shared derived character. We have some rough guidelines, and some seat-of-the-pants feelings, but no unerring formulas. If derived characters are sufficiently "complex," for example, we begin to feel confident that they could not have evolved independently in separate lineages and that their mutual presence therefore indicates common descent.

Chimps and gorillas share a set of complex and apparently independent modifications in several of their chromosomes (mostly "inversions," literally, the turning around of part of a chromosome by breaking, flipping, and reattaching). Since these chromosomal changes are complex and do not seem to represent "easy" modifications so adaptively necessary that separate lineages might evolve them independently, we mark them as shared derived characters present in the common ancestor of chimps and gorillas, and in no other primate. Hence they identify chimps and gorillas as a sister group.

Unfortunately, most derived characters are more ambiguous. They tend to be either easy to construct or else so advantageous that several lineages might evolve them independently by natural selection. Many mammals, for example, develop a sagittal crest—a ridge of bone running along the top of the skull from front to back and serving as an attachment site for muscles. Most primates do not have a sagittal crest, in part because large brains make the cranium bulge and leave neither room nor material for such a structure. But a general rule for scaling of the brain in mammals holds that large animals have relatively smaller brains than relatives of diminished body size (see essays in *Ever Since Darwin* and *The Panda's Thumb*). Thus, the largest primates have a sagittal crest because their relatively small brains do not impede its formation. (This argument does not apply to the great oddball *Homo sapiens,* with an enormous brain despite its large body.) The largest australopithecine, *Australopithecus boisei,* has a pronounced sagittal crest, while smaller members of the same genus do not. Gorillas also have a sagittal crest, while most smaller primates do not. We would make a great error if, using the sagittal crest as a shared derived character, we united an australopithecine with a gorilla in a sister group and linked other, smaller-bodied australopithecines with marmosets, gibbons, and rhesus monkeys. The sagittal crest is a "simple" character, probably part of the potential developmental repertoire for any primate. It comes and goes in evolution, and its mutual presence does not indicate common descent.

Bennett bases her cladistic analysis of the genus *Equus* on skeletal characters, primarily of the skull. All horses are pretty much alike under the skin, and Bennett has not found any shared derived characters as convincing as the chromosomal similarities of chimps and gorillas. Most of her characters are, by her own admission, more like the sagittal crest—hence the provisional nature of her conclusions.

Bennett argues that the genus *Equus* contains two major

cladistic groups—donkeys and asses on one side and true horses and zebras on the other. Thus, zebras pass the first test for consideration as a genealogical unit. Unfortunately (or not, according to your point of view), Bennett claims that they fail the second test. She does identify the Burchell and Grevy zebras (*Equus burchelli* and *E. grevyi*) as a sister group. But in her scheme, the third species, the mountain, or Hartmann, zebra (*E. zebra*) does not join its cousins to form a larger sister group. Instead, the sister species of the mountain zebra is our close compatriot in farm and track, the true horse (*E. caballus*)! Thus, mountain zebras join with true horses before they connect with other zebras. Old Dobbin is inextricably intercalated within the cladogram of zebras—and since he is not a zebra by any definition, then what, if anything, is a zebra?

But Bennett's analysis is based upon only three characters, none very secure. All are potentially simple modifications of shape or proportion, not presences or absences of complex structures. All, like the sagittal crest, could come and go. Only one potential shared derived character unites true horses with *E. zebra:* the "orientation of postorbital bars relative to horizontal plane" (a relatively less slanted position for a bar of bone located on the skull behind the eyes—not exactly the stuff of which confident conclusions are made). Only two potential shared derived characters unite

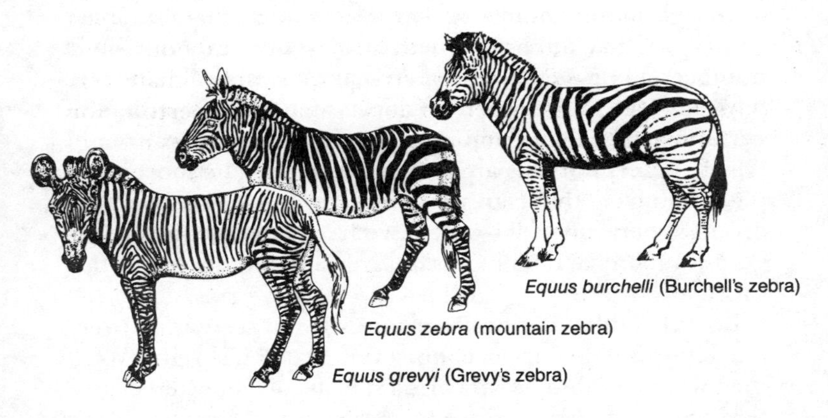

Equus burchelli (Burchell's zebra)

Equus zebra (mountain zebra)

Equus grevyi (Grevy's zebra)

Burchell and Grevy zebras: the presence of frontal doming (inflation of the top part of the skull) and relative skull breadth (these two zebras have long and narrow snouts). Unfortunately, we know that at least one of these characters doesn't work well for Bennett's cladistic scheme because she admits that a member of her other lineage—a horse with the peculiar moniker of the Asiatic half-ass (*E. hemionus*)—has independently evolved a long, narrow snout. If twice, why not three times?

When we look for corroboration to an obvious source—numbers of chromosomes—we are again disappointed. As I discussed in essay 26, the various species of horses, despite their marked similarities of form, differ greatly in number of chromosomes. Fusion or fission of chromosomes may be a major mechanism of speciation in mammals, and these differences may therefore have great evolutionary significance. All zebras, and only zebras, have fewer than fifty pairs of chromosomes (thirty-two in Hartmann's zebra to forty-six in Grevy's zebra). All other horses have more than fifty (from fifty-six in *Equus hemionus* to sixty-six in Przewalski's horse). The low number of zebras might mark them as a genealogical group if the character is shared derived, and not either primitive or evolved more than once. Bennett's hypothesis may still be maintained by arguing either that small numbers are primitive for all horses and that asses and true horses acquired larger numbers by independent evolutionary routes; or that different lineages of zebras evolved small numbers along separate evolutionary paths. Still, since we have no reason to associate stripes with small numbers of chromosomes, their conjoined presence in all zebras might best be interpreted as a sign of genealogy. The more complex characters that a group shares, the more likely that the group is genealogical—unless we have good reason to regard all the characters as primitive (and we do not in this case).

I conclude that Bennett's proposal is interesting, but very much unproven. Suppose, however, that she is right. What then would a zebra be? Or more specifically, how did cladistically unrelated horses get black-and-white stripes? There are two possibilities. Either the common ancestor of zebras and true horses had stripes and true horses lost them, while the three species of "zebras" passively retained them; or else, striping is an inherited developmental capacity of all horses and not so complex a character as it appears. In this case, several separate lineages could acquire stripes independently. Zebras would then be horses that have realized a potential pathway of development probably common to many or all members of the genus *Equus* (see next essay).

This particular tale of zebras may not hold, but the radical messages of cladistic ordering are secure in many cases. Some of our most common and comforting groups no longer exist if classifications must be based on cladograms. With apologies to Mr. Walton and to so many coastal compatriots in New England, I regret to report that there is surely no such thing as a fish. About 20,000 species of vertebrates have scales and fins and live in water, but they do not form a coherent cladistic group. Some—the lungfishes and the coelacanth in particular—are genealogically close to the creatures that crawled out on land to become amphibians, reptiles, birds, and mammals. In a cladistic ordering of trout, lungfish, and any bird or mammal, the lungfish must form a sister group with the sparrow or elephant, leaving the trout in its stream. The characters that form our vernacular concept of "fish" are all shared primitive and do not therefore specify cladistic groups.

At this point, many biologists rebel, and rightly I think. The cladogram of trout, lungfish, and elephant is undoubtedly true as an expression of branching order in time. But must classifications be based only on cladistic information? A coelacanth looks like a fish, tastes like a fish, acts like a fish, and therefore—in some legitimate sense beyond hidebound tradition—*is* a fish.

No debate in evolutionary biology has been more intense during the past decade than the challenge raised by cladistics against traditional schemes of classification. The problem arises from the complexity of the world, not from the fuzziness of human thought (although woolliness has made its usual contribution as well). We must recognize two rather different components to our vernacular conception of "similarity" between organisms—and classifications are designed to reflect relative degrees of similarity. On the one hand, we must consider genealogy, or branching order. Cladistics works with branching order alone, rigorously excluding any other notion of similarity. But what about the admittedly vague and qualitative, but not therefore unimportant, notion of overall similarity in form, function, or biological role? The coelacanth, to say it again, looks and acts like a fish even if its closer cladistic relatives are mammals. Another theory of classification, called phenetics—from a Greek word for appearance—focuses on overall similarity alone and tries to escape the charge of subjectivity by insisting that phenetic classifications be based upon large suites of characters, all expressed numerically and processed by computer.

Unfortunately, these two types of information—branching order and overall similarity—do not always yield congruent results. The cladist rejects overall similarity as a snare and delusion and works with branching order alone. The pheneticist attempts to work with overall similarity alone and tries to measure it in the vain pursuit of objectivity. The traditional systematist tries to balance both kinds of information but often falls into hopeless confusion because they really do conflict. Coelacanths are like mammals by branching order and like trout by biological role. Thus, cladists buy potential objectivity at the price of ignoring biologically important information. And traditionalists curry confusion and subjectivity by trying to balance two legitimate, but often disparate, sources of information. What is to be done?

I cannot answer this question, for it raises issues of style, mores, and methodology more than demonstrable substance. But I can at least comment on the source of this bitter debate—a rather simple point that somehow got lost in the heat. In an ideal world, there would be no conflict among the three schools—cladistics, phenetics, and traditional—and all would produce the same classification for a given set of organisms. In this pipe-dream world, we would find a perfect correlation between phenetic similarity and recency of common ancestry (branching order); that is, the longer ago two groups of organisms separated from a common ancestor, the more unlike they would now be in appearance and biological role. Cladists would establish an order of branching in time by cataloging shared derived characters. Pheneticists would crunch their numerous measures of similarity in their favorite computers and find the same order because the most dissimilar creatures would have the most ancient common ancestors. Traditionalists, finding complete congruence between their two sources of information, would join the chorused harmony of agreement.

But let the reverie halt. The world is much more interesting than ideal. Phenetic similarity often correlates very poorly with recency of common ancestry. Our ideal world requires a constancy of evolutionary rate in all lineages. But rates are enormously variable. Some lineages change not at all for tens of millions of years; others undergo marked alterations in a mere thousand. When the forebears of terrestrial vertebrates first split off from a common ancestry with coelacanths, they were still unambiguously fish in appearance. But they have evolved, along numerous lines during some 250 million years, into frogs, dinosaurs, flamingos, and rhinoceroses. Coelacanths, on the other hand, are still coelacanths. By branching order, the modern coelacanth may be closer to a rhino than a tuna. But while rhinos, on a rapidly evolving line, are now markedly different from that distant common ancestor, coelacanths still look and act like fish—and we might as well say so. Cladists will put them with rhinos, pheneticists with tunas; traditionalists will hone their rhetoric to defend a necessarily subjective decision.

Nature has imposed this conflict upon science by decreeing, through the workings of evolution, such unequal rates of change among lineages and such a poor correlation between phenetic similarity and recency of common ancestry. I do not believe that nature frustrates us by design, but I rejoice in her intransigence nonetheless.

29 How the Zebra Gets Its Stripes

SOME PERSISTENT, unanswered questions about nature possess a kind of majestic intractability. Does the universe have a beginning? How far does it extend? Others refuse to go away because they excite a pedestrian curiosity but seem calculated, in their very formulation, to arouse argument rather than inspire resolution. As a prototype for the second category, I nominate: Is a zebra a white animal with black stripes or a black animal with white stripes? I once learned that the zebra's white underbelly had decided the question in favor of black stripes on a blanched torso. But, to illustrate once again that "facts" cannot be divorced from cultural contexts, I discovered recently that most African peoples regard zebras as black animals with white stripes.

In a poem about monkeys, Marianne Moore discussed some compatriots at the zoo and contrasted elephants and their "strictly practical appendages" with zebras "supreme in their abnormality." Yet we learned in essay 28 that the three species of zebras may not form a group of closest relatives—and that stripes either evolved more than once or represent an ancestral pattern in the progenitors of true horses and zebras. If stripes are not the markers of a few related oddballs but a basic pattern within a large group of animals, then the problems of their construction and meaning acquire more general interest. J. B. L. Bard, an embryologist from Edinburgh, has recently analyzed zebra stripes in the broad context of models for color in all mammals. He detected a developmental unity underlying the different patterns of adult striping among our three species of zebras and, *inter alia,* even proposed an answer to the great black-and-white issue in favor of the African viewpoint.

Biologists follow a number of intellectual styles. Some delight in diversity for its own sake and spend a lifetime describing intricate variations on common themes. Others strive to discover an underlying unity behind the differences that sort these few common themes into more than a million species. Among searchers for unity, the Scottish biologist and classical scholar D'Arcy Wentworth Thompson (1860–1948) occupies a special place. D'Arcy Thompson spent his life outside the mainstream, pursuing his own brand of Platonism and packing insights into his thousand-page classic, *On Growth and Form*—a book so broad in appeal that it won him an honorary degree at Oxford and, thirty years later, entered the *Whole Earth Catalog* as "a paradigm classic."

D'Arcy Thompson struggled to reduce diverse expressions to common generating patterns. He believed that the basic patterns themselves had a kind of Platonic immutability as ideal designs, and that the shapes of organisms could only include a set of constrained variations upon the basic patterns. He developed a theory of "transformed coordinates" to depict variations as expressions of a single pattern, stretched and distorted in various ways. But he worked before computing machines could express such transformations in numerical terms, and his theory achieved little impact because it never progressed much beyond the production of pretty pictures.

As a subtle thinker, D'Arcy Thompson understood that emphases on diversity and unity do not represent different theories of biology, but different aesthetic styles that profoundly influence the practice of science. No student of diversity denies that common generating patterns exist, and no searcher for unity fails to appreciate the uniqueness of particular expressions. But allegiance to one or the other

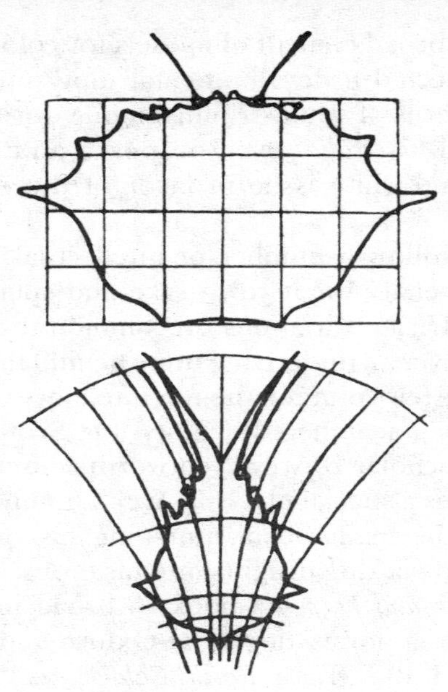

"Transformed coordinates" on the carapaces of crabs from two different genera demonstrate unity of form. FROM ON GROWTH AND FORM, BY D'ARCY WENTWORTH THOMPSON/CAMBRIDGE UNIVERSITY PRESS, 1917.

style dictates, often subtly, how biologists view organisms and what they choose to study. We must reverse the maxim of the reprobate father teaching his son morality—do what I say, not what I do—and recognize that biological allegiances lie not so much in words but in actions and subjects chosen for research. Note what I do, not what I say. Of the "pure taxonomist"—the describer of diversity—D'Arcy Thompson wrote:

> When comparing one organism with another, he describes the differences between them point by point and "character" by "character." If he is from time to time constrained to admit the existence of "correlation" between characters . . . he recognizes this fact of correlation somewhat vaguely, as a phenomenon due

to causes which, except in rare instances, he can hardly hope to trace; and he falls readily into the habit of thinking and talking of evolution as though it had proceeded on the lines of his own description, point by point and character by character.

D'Arcy Thompson recognized, with sadness, that the theme of underlying unity had received much lip service, but little application. Differences between the striping patterns of our three zebra species had been described minutely and much energy had been invested in speculations about the adaptive significance of differences. But few had asked whether all the patterns might be reduced to a single system of generating forces. And few seemed to sense what significance such a proof of unity might possess for the science of organic form.

The vulgar version of Darwinism (not Darwin's) holds that natural selection is so powerful and pervasive in scrutinizing every variation and constructing optimal designs that organisms become collections of perfect parts, each minutely crafted for its special role. While not denying correlation in development or underlying unity in design, the vulgar Darwinian does relegate these concepts to unimportance because natural selection can always break a correlation or remold an inherited design.

The pure vulgar Darwinian may be a fiction; no one could be quite so foolish. But evolutionary biologists have often slipped into the practice of vulgar Darwinism (while denying its precepts) by following the reductionistic research strategy of analyzing organisms part by part and invoking natural selection as a preferred explanation for all forms and functions—the point of D'Arcy Thompson's profound statement cited above. Only in this way can I make sense of the curious fact that unity of design has received so little attention in the practice of research—although much lip service in textbooks—during the past forty years, while evolutionary biologists have generally preferred a rather strict construction of Darwinism in their explanations of nature.

For many reasons, ranging from the probable neutrality

of much genetic variation to the nonadaptive nature of many evolutionary trends, this strict construction is breaking down, and themes of unity are receiving renewed attention. Old ideas are being rediscovered; D'Arcy Thompson, although never out of print, is now often out of bookstores (and in personal libraries). One old and promising theme emphasizes the correlated effects of changes in the timing of events in embryonic development. A small change in timing, perhaps the result of a minor genetic modification, may have profound effects on a suite of adult characters if the change occurs early in embryology and its effects accumulate thereafter.

The theory of human neoteny, often discussed in my essays (see my disquisition on Mickey Mouse in *The Panda's Thumb*) is an expression of this theme. It holds that a slowdown in maturation and rates of development has led to the expression in adult humans of many features generally found in embryos or juvenile stages of other primates. Not all these features need be viewed as direct adaptations built by natural selection. Many, like the "embryonic" distribution of body hair on heads, armpits, and pubic regions, or the preservation of an embryonic membrane, the hymen, through puberty, may be nonadaptive consequences of a basic neoteny that is adaptive for other reasons—the value of slow maturation in a learning animal, for example.

Bard's proposal for "a unity underlying the different zebra striping patterns" follows D'Arcy Thompson's theme of a basic motif stretched and pulled in different ways by varying forces of embryonic growth. These varying forces arise because the basic pattern develops at *different times* in the embryology of the three species. Bard thus combines the theme of transformed coordinates with the insight that substantial evolution can proceed by changes in the timing of development.

The basic pattern is simplicity itself: a series of parallel stripes deposited perpendicular to a line running along the embryonic zebra's back from head to tail—hang a sheet over a taut wire and paint vertical stripes on each side of it. These stripes are initially laid down at a constant size, no matter how big the embryo that forms them. They are 0.4 mm., or approximately 20 cell diameters, apart. The bigger the embryo, the greater the initial number of stripes. (I should point out that Bard's argument is a provocative model for testing, not a set of observations; no one has ever traced the embryology of zebra striping directly.)

The three zebra species differ in both number and configuration of stripes. In Bard's hypothesis, these complex variations arise only because the same basic pattern—the parallel stripes of constant spacing—develops during the fifth week of embryonic growth in one species, during the fourth week in another, and during the third week in the third species. Since the embryo undergoes complex changes in form during these weeks, the basic pattern is stretched and distorted in varying ways, leading to all the major differences in adult striping.

The three species differ most notably in patterns of striping on the rump and hind quarters (see illustration in the last chapter). Grevy's zebra *(Equus grevyi)* has numerous fine and basically parallel stripes in these rear regions. On Bard's model, the stripes must have formed when the back part of the embryo was relatively large. (The larger the part, the more stripes it receives, since stripes are initially formed at constant size and spacing.) In the embryology of horses, the tail and hind regions expand markedly during the fifth week *in utero.* If adults possess numerous, fine posterior stripes, they must form after this embryonic expansion of the rear quarters. (Unfortunately, no one has ever studied the early embryology of zebras directly, and Bard assumes that the intrauterine growth patterns of true horses are followed by their striped relatives as well. Since basic features of early embryology tend to be highly conservative in evolution, true horses are probably fair models for zebras.)

The mountain zebra, *Equus zebra,* looks much like *E. grevyi* until we reach the haunch, where three broad stripes substitute for the numerous fine stripes of Grevy's zebra. Broad stripes on adults indicate initial formation on a small piece of embryo (where few stripes could fit), and later rapid growth of the piece (widening the stripes as the general area expands). If an embryo forms stripes in its fourth week, just before the posterior expansion that provides room for the many fine marks of Grevy's zebra, it will build the pattern of a mountain zebra during later embryonic growth.

Burchell's zebra, *Equus burchelli,* also has just a few broad stripes on its haunch. But, while the mountain zebra has fine stripes over most of its back and broad stripes only over the haunch, the broad stripes of Burchell's zebra begin in the middle of the belly and sweep back over the haunch. This pattern suggests an initial formation of stripes during the third week of embryonic growth. At this early stage, the embryo has a short, compact back, which later expands toward the rear in a broad, arching curve while the belly remains short. A stripe that initially ran vertically from belly to spine would be pulled toward the rear as the embryo's top surface expanded backward while its belly grew little. An adult stripe, subject to such deformation in its embryonic life, would be broad and would run from the belly up and over the haunch—as in Burchell's zebra.

Thus, Bard can render differences in rear striping of all three species as the results of deforming the same initial pattern at different times during normal embryonic growth. His hypothesis receives striking support from another source: the total number of stripes itself. Remember that Bard assumes a common size and spacing for stripes at their initial formation. Thus, the larger the embryo when stripes first form, the greater the number of stripes. Grevy's zebra, presumably forming its stripes as an embryo of five weeks and about 32 mm. in length, has eighty or so stripes as an adult—or about 0.4 mm. per stripe. Mountain zebras with a fourth-week embryo of some 14 to 19 mm. have about forty-three stripes—again about 0.4 mm. per stripe. Burchell's zebra has twenty-five to thirty stripes; if they form in a third-week embryo some 11 mm. long, we get the same value—about 0.4 mm. per stripe.

As additional support, and a lovely example of the difference between superficial appearance and knowledge of underlying causes, consider an old paradox involving hybrid offspring between zebras and true horses. These animals almost always have more stripes than their zebra parent. "Common sense," based on superficial appearance, declares this result puzzling. After all, the state between stripes and no stripes is few stripes. But if Bard is right about the underlying causes of striping, then this paradoxical result makes sense. The intermediate state between stripes and no stripes might well be a *delay* in the embryonic formation of stripes. If stripes then begin at their common size and spacing upon a larger embryo, the resulting adult will have *more* stripes.

If a unity of basic architecture underlies the diversity of zebra striping, then we must suspect that we are confronting a general pattern in nature, not just the "supreme abnormality" that Marianne Moore described. Darwin viewed horses in this light and recognized that the capacity for striping in all horses constituted a powerful argument for evolution itself. If zebras are odd and perfect adaptations for camouflage, God might have made them as we find them. But if zebras merely actuate and exaggerate a potential property of all horses, then the occasional realization of striping in other horses—where it cannot be viewed as a perfected adaptation ordained by God—must indicate a community of evolutionary descent.

Darwin devoted much space in chapter 5 of the *Origin of Species* to an exhaustive tabulation of occasional striping in other horses. Asses, he found, often have "very distinct transverse bars . . . like those on the legs of a zebra." True horses often possess a spinal stripe, and some also have transverse leg bars. Darwin found a Welsh pony with three parallel stripes on each shoulder. And he noted that hybrids (with no zebra parents) were often strongly striped—an example of the common, and still mysterious, observation that hybrids often display ancestral reminiscences present in neither parent. "I once saw a mule," Darwin wrote, "with its legs so much striped that any one at first would have thought that it must have been the product of a zebra."

From this illustration of common, and often nonadaptive,

patterns in all horses, Darwin drew one of his most powerful and passionate arguments for evolution—well worth quoting *in extenso:*

> He who believes that each equine species was independently created, will, I presume, assert that each species has been created with a tendency to vary, both under nature and under domestication, in this particular manner, so as often to become striped like other species of the genus; and that each has been created with a strong tendency, when crossed with species inhabiting distant quarters of the world, to produce hybrids resembling in their stripes, not their own parents, but other species of the genus. To admit this view is, as it seems to me, to reject a real for an unreal, or at least for an unknown, cause. It makes the works of God a mere mockery and deception; I would almost as soon believe with the old and ignorant cosmogonists, that fossil shells had never lived, but had been created in stone so as to mock the shells now living on the seashore.

The same theme also suggests an answer to the title of essay 28: "What, If Anything, Is a Zebra?" I advanced the argument that zebras may not form a group of closest relatives but a set of different horses that had either evolved stripes independently or inherited them from a common ancestor (while asses and true horses lost them). Bard's hypothesis lends support to this conjecture because it suggests that the underlying pattern of zebra striping may be so simple that all horses include it in their repertoire of development. Zebras, then, may be the realization of a potential possessed by all horses.

Finally, moving from the sublime to the merely interesting, Bard proposes a solution to the primal dilemma and argues that zebras are black animals with white stripes after all. The white underbelly, he points out, is a lousy argument because many fully colored mammals are white underneath. Color may be generally inhibited in this region for reasons at present unknown. Mammals do not have their colors painted on a white background. The basic issue may then be rephrased: Does striping result from an inhibition or a deposition of melanin? If the first, zebras are black animals; if the second, they are white with black stripes.

Biologists often look to teratologies, or abnormalities of development, to solve such issues. Bard has uncovered an abnormal zebra whose "stripes" are rows of dots and discontinuous blotches, rather than coherent lines of color. The dots and blotches are white on a black background. Bard writes: "It is only possible to understand this pattern if the white stripes had failed to form properly and that therefore the 'default' color is black. The role of the striping mechanism is thus to inhibit natural pigment formation rather than to stimulate it." The zebra, in other words, is a black animal with white stripes.

30 Quaggas, Coiled Oysters, and Flimsy Facts

AS DARWIN CATALOGED cases of striped horses and asses to illustrate a common ancestry with zebras (see last essay), he inevitably encountered one of the most famous animals of nineteenth-century natural history—the Earl of Morton's mare. Darwin wrote in the *Origin of Species:* "In Lord Moreton's [*sic*] famous hybrid from a chestnut mare and male quagga, the hybrid, and even the pure offspring subsequently produced from the mare by a black Arabian sire, were much more plainly barred across the legs than is even the pure quagga."

The quagga, a zebra with stripes restricted to its neck and forequarters, is now extinct. It was not thriving in the early nineteenth century either, when the good Earl hoped to save the species by domesticating it. He was able to procure a male for his noble experiment but could never obtain a female. So he bred his male quagga with "a young chestnut mare of seven-eighths Arabian blood," and obtained a hybrid with "very decided indications of her mixed origin." Nothing surprising so far.

But the disappointed Lord Morton, unable to find more quaggas, sold his Arabian mare to Sir Gore Ouseley, who proceeded to breed her with "a very fine black Arabian horse." When Lord Morton visited his friend and viewed the two offspring of pedigreed Arabian parents, he was astonished to note in them what he took to be a "striking resemblance to the quagga." Somehow, the quagga father had influenced subsequent offspring sired by other males years after his permanent departure from the life of Lord Morton's mare. How could such an influence be maintained long after physical contact had ceased?

Lord Morton's mare was the most celebrated, but by no means the only, case of a phenomenon that German biologist August Weismann later named "telegony," from Greek roots meaning "offspring at a distance," or the idea that sires could influence subsequent progeny not fathered by them. Since the supposed phenomenon turned out to be an illusion, telegony now reposes as one more forgotten item on history's ash heap, and neither Lord Morton nor his mare retain any notoriety today.

But historian of science Richard W. Burkhardt, Jr., who recently wrote an excellent article on the history of telegony in general and Lord Morton's mare in particular (see bibliography), has demonstrated that telegony was once a major subject for research and "inspired the most extensive work in experimental animal breeding conducted in Britain between Darwin's death in 1882 and the rediscovery of Mendel's law in 1900." Darwin himself was a major supporter of telegony.

If the supposed causes of telegony are a bit mysterious, Darwin's reasons for espousing the idea may seem equally difficult to fathom. After all, he first discussed the progeny of Lord Morton's mare in a context that implied an explanation opposed to telegony. As I wrote in the first two parts of this trilogy, Darwin had cataloged all cases he could find of asses and true horses with stripes. He used these striped horses as an effective argument for evolution: If God had created true horses, asses, and zebras as separate forms, why should we find occasional striped individuals in species that normally lack them? Does not this latent tendency for striping in all horses (permanently actualized only in zebras) indicate common descent? Why, then, did Darwin later implicate the previous quagga sire in the striping of subsequent offspring from Lord Morton's mare? In this first discussion, from the *Origin of Species,* Darwin set out to prove that true horses and asses can develop stripes *without* any zebra influence. As we shall see, this original explanation was apparently correct.

Burkhardt argues that Darwin changed his mind and supported telegony because it fit so well with the (ironically) "non-Darwinian" theory of heredity that he developed in 1868. Under this "provisional hypothesis of pangenesis" (as Darwin called it), all cells of the body produce tiny particles, called gemmules, that course throughout the body, gather in the sex cells, and eventually transmit the characters of parents to offspring. Since gemmules might be altered if the cells producing them are changed by the influence of environment or the activity of animals themselves, acquired characters can be inherited and evolution has an important Lamarckian aspect.

Telegony meshed well with pangenesis because gemmules included with the quagga's sperm would have remained in the body of Lord Morton's mare and extended their influence to her subsequent offspring. (Darwin even once speculated that gemmules passed in sperm might explain why some women grow to resemble their husbands. As to why so many people resemble their dogs, Darwin maintained a discreet silence.)

Telegony finally fell when a new theory of heredity rose to favor and excluded it. August Weismann, who championed the strict Darwinism of natural selection against all forms of Lamarckian inheritance (including Darwin's own pangenesis), argued for what he called "the continuity of germ plasm." He held that reproductive cells are completely isolated from the rest of the body and cannot be influenced by whatever forces mold and alter other organs and tissues. Acquired characters cannot affect the next generation because they cannot penetrate the "casket" that holds reproductive cells and transmits itself in toto from generation to generation. (The fertilized egg, of course, is formed by the union of two reproductive cells. As it starts to divide, however, the nonreproductive cells form, eventually develop into the organism's body, and become rigidly segregated from the continuous lineage of germ cells. Telegony makes no sense because, even if they existed, male gemmules in a female's body could not get to the germ cells—unless they managed to reach the ovaries themselves and modify the immature ova.)

Telegony kicked about in scientific literature for seventy years, from Morton's note to the Royal Society in 1820 until Weismann's challenge. When Weismann proposed the continuity of germ plasm, telegony became a threatening anomaly demanding affirmation or rejection. Many tests were made, and telegony failed them miserably. In particular, J. C. Ewart, Regius Professor of Natural History at Edinburgh, tried to repeat Morton's own experiment. Since quaggas had joined *Eohippus* in the realm of departed horses, Ewart mated twenty mares of different races and breeds with a male Burchell's zebra. The first hybrid, born in 1896, had stripes as expected. Ewart then mated the mare with a second sire, an Arab stallion. Their offspring also had stripes, albeit faint, and telegony seemed vindicated. But Ewart knew that he needed controls and therefore bred the same Arab stallion to other mares that "had never so much as seen a zebra." The offspring of these matings were as richly striped as the foal from the mare that had previously mated with a zebra. Darwin had been right the first time. Stripes do not arise from a mysterious previous zebra influence; they represent a potential pathway of development for all horses.

I have not recounted this tale of telegony for its own sake, since antiquarian musings only excite professionals and trivia buffs. Rather, as Burkhardt emphasizes, the story embodies a larger, troubling, and important issue about the nature of fact in science. Telegony, so far as we can tell, was wrong; yet it remained in the literature as a pristine fact, largely unchallenged, for seventy years. In a reversal of the stereotyped scenario, where a single sturdy fact arises to destroy an entire edifice of theory, the "fact" of telegony came first, became entrenched, and was only seriously challenged when a theory—Weismann's continuity of germ plasm—rendered it anomalous. Burkhardt notes that in the usual model "a long-accepted theory is toppled by a newly discovered, apparently anomalous fact. In the case of telegony, in contrast, a long-accepted 'fact' was discredited when confronted with a new, apparently contradictory theory."

In part, Burkhardt suggests, telegony gathered favor because it fit well with a variety of nineteenth-century assumptions, ranging from the "natural" dominance of males over females to support for separation of races by the argument that sexual contact with a lower race might extend its baleful influence far beyond the immediate consequences of the act itself. In part, telegony simply wasn't controversial enough and nobody bothered to test Lord Morton's improbable assertion. When tested, telegony fell, and to this extent the usual model of science as objective experiment was vindicated. But another aspect of the conventional view did not apply: fact did not act as the cleansing broom for outmoded theory. Rather, a false "fact" endured for an uncomfortably long time until theory demanded its test. What does this story tell us about the relation of fact and theory in science and about the role of single, isolated facts in the first place? I will return to these generalities, but first, another and similar tale.

In 1922, British paleontologist A. E. Trueman published the most famous paper of our century on a supposedly unbroken lineage of evolving fossils. He argued that flat oysters had slowly evolved into coiled oysters of the genus *Gryphaea.* Although coiling was originally advantageous in raising the oysters above an increasingly muddy sea floor, the trend, once started, could not be halted. *Gryphaea* built one coiled valve and one flat valve lying like a cap atop its coiling partner. Eventually, the coil grew up and over the cap, finally pressing hard upon it. Unable to open its shell, *Gryphaea* perished, imprisoned in its own embrace.

When Trueman published his paper, most paleontologists were not Darwinians. The theory of orthogenesis, or "straight line" evolution forcing organisms along predetermined paths, was still popular. An inexorable trend that natural selection could not stop and that led to the demise of a lineage was an expected phenomenon. Thus, Trueman's story was not challenged by paleontologists, and like the zebra's impact on Lord Morton's mare, overcoiled *Gry-*

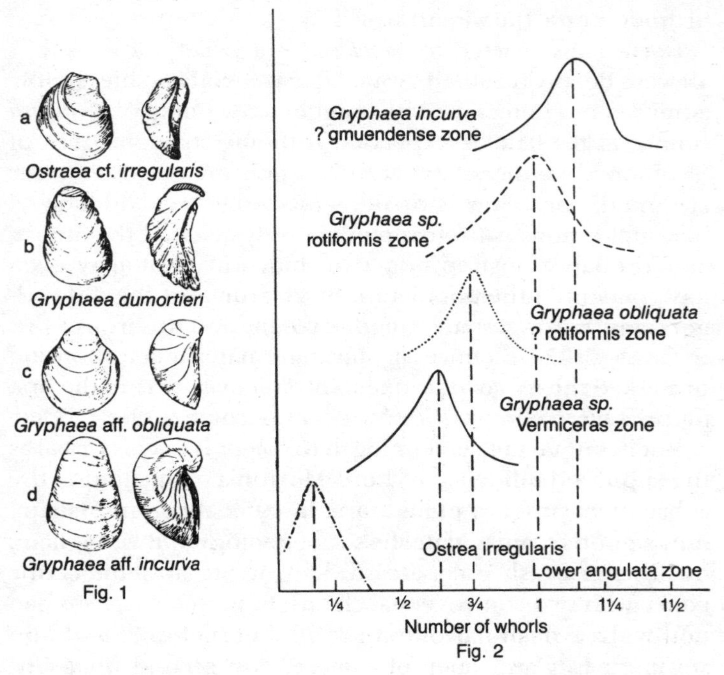

Supposed increase of coiling in *Gryphaea* according to Trueman. Ancestors (top) to coiled descendants (bottom) in pictures of oysters to left. Ancestors (bottom) to coiled descendants (top) in charts of variation in number of coils to right. FROM K. JOYSEY, BIOLOGICAL REVIEWS, VOL. 34, 1959.

phaea became an established fact.

In the late 1930s, after a gestation of nearly one hundred years from Darwin's great insight and eighty years from its publication, natural selection finally triumphed as the accepted theory of evolutionary change, and *Gryphaea* became an anomaly. How could a lineage *actively* evolve itself to extinction if evolutionary change is directed by natural selection and can, therefore, only occur in directions that adapt organisms to local environments? (Extinction based upon an inability to change rapidly enough in the face of environmental perturbation is another matter and perfectly orthodox in a Darwinian world.)

Darwinians reacted to *Gryphaea* in a variety of ways, but all were deeply troubled. Some, like J. B. S. Haldane, simply admitted puzzlement: "The exaggerated coiling of *Gryphaea* cannot at present be explained with any strong degree of likelihood." Others, like G. G. Simpson, tried to steer around the problem with faintly plausible, but evidently ad hoc suggestions: since overcoiling only affected the oldest, and probably postreproductive, individuals, it may even have benefited the population by clearing out the old codgers and making space for the young and vigorous. Yet, through all these efforts in salvaging natural selection, no one asked the more basic question: Is it even true in the first place? Overcoiling in *Gryphaea* had become a fact.

But it is not true; overcoiling in *Gryphaea* is as chimerical as the supposed influence of Lord Morton's quagga upon the subsequent progeny of his mare. In 1959, Anthony Hallam, now a good friend and professor of geology at Birmingham, but then a brash graduate student, wrote an iconoclastic paper with two controversial claims: first, that *Gryphaea* had not evolved from flat oysters at all, but had migrated into southern England from elsewhere; and second, that *Gryphaea* arrived in England as coiled as it would ever be—and that no trend to increased coiling within *Gryphaea* could be demonstrated at all. Hallam's conclusion scandalized many senior scientists who had known Trueman. H. H. Swinnerton, dean of British paleontology, wrote back in anger, accusing Hallam of both "sin" and "monstrous error," an odd pair of accusations for a scientific paper.

Twenty years later the dust has settled, and I do not think that anyone would now doubt Hallam's debunking. For Hallam's first claim—that *Gryphaea* did not evolve from flat oysters—H. B. Stenzel's impressive monograph (*Treatise on Invertebrate Paleontology,* part N, 1971) has proved that the lineages of flat oysters (genus *Ostrea*) and coiled oysters (genus *Gryphaea*) have been separate from their first appearance, that *Gryphaea* is a bit older than *Ostrea,* and that coiled ancestors for Trueman's *Gryphaea* were living in Greenland before *Ostrea* ever appeared in England.

For his second claim—that *Gryphaea* did not increase in coiling during its evolution—I have reexamined all the published data and concluded that while *Gryphaea* did increase in body size, its degree of adult coiling remained constant. Ironically, this means that later *Gryphaea* were actually more loosely coiled than earlier specimens of the same body size, for coiling increases with growth, and if a larger adult descendant will reach the same intensity of coiling as its smaller adult ancestor, then the descendant's shell must be more loosely coiled when it is still a juvenile of the same size as an ancestral adult. (I have reprinted all major papers from the great *Gryphaea* debate in a volume entitled: *The Evolution of Gryphaea,* Arno Press, 1980.)

If Trueman's edifice toppled so easily, why was his "fact" accepted so readily in the first place? One might suspect that the overcoiling of *Gryphaea* had received lengthy and complex documentation, however unreliable it proved to be, and that Trueman simply snowed potential critics. Not at all. The original evidence was flimsy almost beyond belief. In his 1922 paper, Trueman showed how most large adult *Gryphaea avoid* calamitous overcoiling by decreasing the tightness of their spiral late in growth.

Trueman's claim for overcoiling was based on a single specimen, the "type" (or name-bearing) specimen of the species *Gryphaea incurva.* I examined this specimen at the British Museum in 1971. At first glance, the coiled valve does seem to press hard upon the flat valve. But the specimen was found in very fine-grained muds of the same color as the shell itself. A bit of probing revealed—and X-ray photographs later confirmed—that the supposed "lock" upon the flat valve had not been formed by the coiled valve itself, but by mud that wedged its way into the space between flat and coiled valve—the space that allowed the shell to open—after the animal's death.

If telegony and overcoiling are false "facts," why did each command prestige and inspire no attempt at refutation for so long? I believe, first of all, that the reputation of false facts is bolstered by the naïve belief that facts are bits of unsullied information extracted from nature by an objective process of pure observation or by scientific inference. But facts arise in contexts of expectation, and both the eye and the mind are notoriously fallible instruments. (Anyone who thinks that claims for direct observation possess some special, irrefutable status should read Elizabeth Loftis's chilling book on eyewitness testimony.) Paleontologists, accepting the reality of orthogenesis, were primed to believe in *Gryphaea*'s overcoiling; telegony seemed reasonable until Weismann challenged it seventy years later. Secondly, facts achieve an almost immortal status once they pass from primary documentation into secondary sources, particularly textbooks. No publication is quite so conservative as a textbook; errors are copied from generation to generation and seem to gain support by sheer repetition. No one goes back to discover the fragility of original arguments.

I am not trying to convey the message that all knowledge is relative and that facts can never achieve universal approbation—quite the reverse. Rather, we have to distinguish between the kinds of factual claims that can achieve acceptance and those that must remain in limbo. The most troublesome facts are single cases—the offspring of Lord Morton's mare, the overcoiling of one *Gryphaea.* We must, as William Bateson advised, "treasure our exceptions." But we must also be aware that single cases are fragile, and that sturdy facts are pervasive patterns in nature, not individual peculiarities. Most "classic stories" in science are wrong.

The need to distinguish sturdy fact (pervasive pattern) from shaky factual claim (single cases with dubious documentation) has never been more evident to me than in the current debate between evolutionists and so-called "scientific creationists." The fact of evolution is as sturdy as any claim in science. Its sturdiness resides in a pervasive pattern detected by several disciplines—for example, the age of the earth and life as affirmed by astronomy and geology, and the pattern of imperfections in organisms that record a history of physical descent.

Against this pattern, creationists employ a destructive, shotgun approach. They present no testable alternative but fire a volley of rhetorical criticism in the form of unconnected, shaky factual claims—a potpourri (literally, a rotten pot, in this case) of nonsense that beguiles many people because it masquerades in the guise of fact and trades upon the false prestige of supposedly pure observation.

The individual claims are easy enough to refute with a bit of research. Creationists themselves have been forced to retreat from the more embarrassing items. Noted creationist Henry Morris, for example, has often cited the supposed footprints of dinosaurs and humans together in rocks of the Paluxy River of Texas. But creationist Leonard Brand attributes some of the "human" prints to erosion and others to a three-toed dinosaur. He also adds: "We do know that there was a fellow during the Depression who carved tracks."

Yet each time we explode one creationist "fact," two more are invented to take its place. Hercules finally killed the Lernaean Hydra, a beast with similar tendencies toward proliferation after partial destruction. We can deprive creationism of all intellectual respectability (though not, alas, of some popular appeal) by remembering that sturdy facts are built from widespread patterns and that coherence in structure is the sign of strong arguments and theories. Unconnected, individual items remain shaky until they form a pattern or attain a confidence in individual documentation that neither telegony nor overcoiling—not to mention any creationist claim—ever achieved.

If shaky factual claims were always easy to dislodge, this column could end on a purely optimistic note. But telegony lasted for seventy years, and the ghost of William Jennings Bryan again stalks our nation. If I end with measured optimism, however, I do so in urging that we focus upon the second phrase of what may be Darwin's most famous statement (from the *Descent of Man*): "False facts are highly injurious to the progress of science, for they often endure long; but false views, if supported by some evidence, do little harm, for every one takes delight in proving their falseness."

Bibliography

Adams, M. B. 1978. Nikolay Ivanovich Vavilov. *Dictionary of Scientific Biography* 15:505–13. New York: Charles Scribner's Sons.

Adler, J. 1976. The sensing of chemicals by bacteria. *Scientific American* 234 (4):40–47.

Agassiz, E. C. 1895. *Louis Agassiz: his life and correspondence.* Boston: Houghton, Mifflin.

Agassiz, L. 1874. Evolution and permanence of type. *Atlantic Monthly.*

Agassiz, L. 1885. *Geological sketches.* Boston: Houghton, Mifflin (reprint of essays, mostly from the 1860s).

Alcock, J. 1975. *Animal behavior, an evolutionary approach.* Sunderland, MA: Sinauer Associates.

Alvarez, L. W.; W. Alvarez; F. Asaro; and H. V. Michel. 1980. Extraterrestrial cause for the Cretaceous-Tertiary extinction. *Science* 208:1095–1108.

Aristotle. 1960 ed. *Organon* (Posterior analytics) translated by H. Tredennick. Loeb Classical Library Number 391. Cambridge, MA: Harvard University Press.

Aristotle. 1965 ed. *Historia animalium,* 3 volumes, translated by A. L. Peck. Loeb Classical Library Numbers 437–439. Cambridge, MA: Harvard University Press.

Barash, D.P. 1976. Male response to apparent female adultery in the mountain bluebird: an evolutionary interpretation. *American Naturalist* 110:1097–1101.

Bard, J. B. L. 1977. A unity underlying the different zebra striping patterns. *Journal of Zoology* (*London*) 183:527–39.
Bateson, W. 1894. *Materials for the study of variation.* London: MacMillan.
Bennett, D. K. 1980. Stripes do not a zebra make. *Systematic Zoology* 29:272–87.
Berg, H.C. 1975. How bacteria swim. *Scientific American* 229 (6): 24–37.
Berg, H. C., and R. A. Anderson. 1973. Bacteria swim by rotating their flagellar filaments. *Nature* 245:380–92.
Boule, M. 1921. *Les hommes fossiles.* Paris: Masson.
Britten, R. J., and E. H. Davidson. 1971. Repetitive and non-repetitive DNA sequences and a speculation on the origins of evolutionary novelty. *Quarterly Review of Biology* 46:111–33.
Buckland, W. 1836. *Geology and mineralogy considered with reference to natural theology.* 1841 ed. Philadelphia: Lea and Blanchard.
Bulliet, R. W. 1975. *The camel and the wheel.* Cambridge, MA: Harvard University Press.
Burkhardt, R. W., Jr. 1979. Closing the door on Lord Morton's mare: the rise and fall of telegony. *Studies in the History of Biology* 3:1–21.
Chase, A. 1977. *The legacy of Malthus.* New York: A. Knopf.
Costello, P. 1981. Teilhard and the Piltdown hoax. *Antiquity* 45:58–59.
Crick, F. 1981. *Life itself.* New York: Simon and Schuster.
Cuénot, C. 1965. *Teilhard de Chardin, a biographical study.* Baltimore: Helicon.
Cuvier, G. 1812. *Recherches sur les ossemens fossiles des quadrupèdes,* 4 volumes. Paris: Deterville.
Cuvier, G. 1817. *Essays on the theory of the earth.* With geological illustrations, by Professor Jameson. Edinburgh.
Darwin, C. 1842. *The structure and distribution of coral reefs.* London: Smith, Elder.
Darwin, C. 1859. *On the origin of species by means of natural selection.* London: John Murray.
Darwin, C. 1862. *On the various contrivances by which British and*

foreign orchids are fertilized by insects. London: John Murray.

Darwin, C. 1871. *The descent of man and selection in relation to sex.* London: John Murray.

Darwin, C. 1881. *The formation of vegetable mould, through the action of worms.* London: John Murray.

Davies, G. L. 1969. *The earth in decay, a history of British geomorphology 1578–1878.* New York: American Elsevier.

Dawkins, R. 1976. *The selfish gene.* New York: Oxford University Press.

Doolittle, W. F., and C. Sapienza. 1980. Selfish genes, the phenotype paradigm, and geome evolution. *Nature* 284:-601–603.

Eldredge, N., and S. J. Gould. 1972. Punctuated equilibria: An alternative to phyletic gradualism. In: T. J. M. Schopf, ed., *Models in Paleobiology.* San Francisco: Freeman, Cooper and Co., pp. 82–115.

Fabre, J. H. 1901. *Insect life.* London: MacMillan.

Fabre, J. H. 1918. *The wonders of instinct.* London: T. Fisher Unwin Ltd.

Geikie, A. 1905. *The founders of geology.* London: MacMillan.

Ghiselin, M. 1974. *The economy of nature and the evolution of sex.* Berkeley: University of California Press.

Ginger, R. 1958. *Six days or forever?* Boston: Beacon Press.

Gish, D. T. 1979. *Evolution? The fossils say no!* San Diego: Creation-Life Publishers.

Goddard, H. H. 1913. The Binet tests in relation to immigration. *Journal of Psycho-Asthenics* 18:105–107.

Goddard, H. H. 1917. Mental tests and the immigrant. *Journal of Delinquency* 2:243–77.

Goldschmidt, R. 1940. *The material basis of evolution.* New Haven: Yale University Press (reprinted 1982, with introduction by S. J. Gould).

Gould, S. J. 1972. Allometric fallacies and the evolution of *Gryphaea:* A new interpretation based on White's criterion of geometric similarity. In: T. Dobzhansky, et al., eds., *Evolutionary Biology,* volume 6, pp. 91–118.

Gould, S. J. 1977. *Ever Since Darwin.* New York: W. W. Norton.

Gould, S. J. 1980. *The Panda's Thumb.* New York: W. W. Norton.

Gould, S. J., and R. C. Lewontin. 1979. The spandrels of San Marco and the Panglossian paradigm: a critique of the adaptationist programme. *Proceedings of the Royal Society, London* B205:581–98.

Gould, S. J., and Elisabeth S. Vrba. 1982. Exaptation—a missing term in the science of form. *Paleobiology* 8 (1): 4–15.

Grabiner, J. V., and P. D. Miller. 1974. Effects of the Scopes Trial. *Science* 185:832–37.

Hallam, A. 1959. On the supposed evolution of *Gryphaea* in the Lias. *Geological Magazine* 96:99–108.

Hampé, A. 1959. Contribution à l'étude du développement et de la régulation des déficiences et des excédents dans la patte de l'embryon de poulet. *Archives d'anatomie et de microscopie morphologique* 48:345–478.

Harrison Matthews, L. 1939. Reproduction in the spotted hyena *Crocuta crocuta* (Erxleben). *Philosophical transactions of the Royal Society,* Series B 230:1–78.

Harrison Matthews, L. 1981. Piltdown man—the missing links. Serialized weekly in *New Scientist,* April 30–July 2.

Hutton, J. 1788. Theory of the earth; or an investigation of the laws observable in the composition, dissolution, and restoration of land upon the globe. *Transactions of the Royal Society of Edinburgh* 1:209–304.

Hutton, J. 1795. *Theory of the earth, with proofs and illustrations.* Edinburgh.

Huxley, J. 1942. *Evolution, the modern synthesis.* London: George Allen and Unwin.

Huxley, T. H. 1893. Evolution and ethics in *Evolution and ethics and other essays.* Volume 9 (published 1894) of T. H. Huxley's Collected Essays. New York: D. Appleton and Company.

Jarvis, E. 1842. Statistics of insanity in the United States. *Boston Medical and Surgical Journal* 27:116–21 and 281–82.

Jarvis, E. 1844. Insanity among the colored population of

the free states. *American Journal of the Medical Sciences,* New Series 7:71–83.

Kirby, W., and W. Spence. 1856. *An introduction to entomology* (7th ed.). London: Longman, Green, Longman, Roberts and Green.

Kollar, E. J., and C. Fisher. 1980. Tooth induction in chick epithelium: expression of quiescent genes for enamel synthesis. *Science* 207:993–95.

Kruuk, H. 1972. *The spotted hyena.* Chicago: University of Chicago Press.

Lewis, E. B. 1978. A gene controlling segmentation in *Drosophila. Nature* 276:565–70.

Lukas, M. 1981a. Teilhard and the Piltdown hoax: A playful prank gone too far? Or a deliberate scientific forgery? Or, as it now appears, nothing at all? *America,* May 23, 424–27.

Lukas, M. 1981b. Gould and Teilhard's "fatal error." *Teilhard Newsletter* 14:4–6.

Lurie, E. 1960. *Louis Agassiz, a life in science.* Chicago: University of Chicago Press.

Lyell, C. 1830–1833. *Principles of geology,* 3 volumes. London: John Murray.

Lysenko, T. D. 1954. *Agrobiology, essays on problems of genetics, plant breeding and seed growing.* Moscow: Foreign Languages Publishing House (reprints all articles quoted in essay on Vavilov).

Marsh, O. C. 1892. Recent polydactyle horses. *American Journal of Science* 43:339–55.

Marshall, L. G.; S. D. Webb; J. J. Sepkoski, Jr.; and D. M. Raup. Mammalian evolution and the great American interchange. *Science* 215:1351–57.

McPhee, J. 1981. *Basin and range.* New York: Farrar, Straus and Giroux.

Mivart, St. G. 1871. *On the genesis of species.* London: MacMillan.

Moon, T. J.; P. B. Mann; and J. H. Otto. 1956. *Modern Biology.* New York: Henry Holt and Company.

Nelkin, D. 1977. *Science textbook controversies and the politics of equal time.* Cambridge, MA: Massachusetts Institute of Technology Press.

Nelson, J. B. 1968. *Galápagos: Islands of birds.* London.

Nelson, J. B. 1978. *The Sulidae—gannets and boobies.* Oxford: Oxford University Press.

Ohno, S. 1970. *Evolution by gene duplication.* New York: Springer

Oliver, J. H., Jr. 1962. A mite parasitic in the cocoons of earthworms. *Journal of Parasitology* 48:120–23.

Orgel, L. E., and F. H. C. Crick. 1980. Selfish DNA: the ultimate parasite. *Nature* 284:604–7.

Ouweneel, W. J. 1976. Developmental genetics of homoeosis. *Advances in Genetics* 18:179–248.

Pearson, K., and M. Moul. 1925. The problem of alien immigration into Great Britain, illustrated by an examination of Russian and Polish Jewish children. *Annals of Eugenics* 1:5–127.

Pietsch, T. W. 1976. Dimorphism, parasitism, and sex: reproductive strategies among deep sea ceratioid anglerfishes. *Copeia* No. 4:781–93.

Quinn, T. C., and G. B. Craig, Jr. 1971. Phenogenetics of the homeotic mutant proboscipedia in *Aedes albopictus. Journal of Heredity* 62:1–12.

Racey, P. A., and J. D. Skinner. 1979. Endocrine aspects of sexual mimicry in spotted hyenas *Crocuta crocuta. Journal of Zoology, London* 187:315–26.

Ralls, K. 1976. Mammals in which females are larger than males. *Quarterly Review of Biology* 51: 245–76.

Raup, D. M. 1979. Size of the Permo-Triassic bottleneck and its evolutionary implications. *Science* 206:217–18.

Raup, D. M., and J. J. Sepkosi, Jr. 1982. Mass extinctions in the marine fossil record. *Science* 215:1501–3.

Regan, C. T. 1925. Dwarfed males parasitic on the females in oceanic anglerfishes (Pediculati: Ceratioidea). *Proceedings of the Royal Society* Series B 97:386–400.

Regan, C. T. 1926. The pediculate fishes of the suborder Ceratioidea. Dana Reports, Volume 2.

Schmitz-Moormann, K. (ed.). 1960– . *Pierre Teilhard de Char-*

din—l'oeuvre scientifique (a facsimile edition of Teilhard's scientific works in 14 volumes). Olten and Freiburg im Breisgau: Walter-Verlag.

Schmitz-Moormann, K. 1981. Teilhard and the Piltdown hoax. *Teilhard Newsletter* 14:2–4.

Scopes, J. 1967. *Center of the storm.* New York: Holt, Rinehart and Winston.

Simmons, K. E. L. 1970. Ecological determinants of breeding adaptations and social behavior in two fish-eating birds. In: J. H. Crook (ed.), *Social behavior in birds and mammals.* London: Academic Press.

Stanton, W. 1960. *The leopard's spots: scientific attitudes towards race in America 1815–1859.* Chicago: University of Chicago Press.

Steno, N. 1669. *De solido intra solidum naturaliter contento dissertationis prodromus.* Translated by J. G. Winter, 1916, as The prodromus of Nicolaus Steno's dissertation. New York: Macmillan.

Stenzel, H. B. 1971. Oysters. Treatise on Invertebrate Paleontology Part N, Volume 3, Mollusca 6, Bivalvia. Geological Society of America and the University of Kansas.

Struhl, G. 1981. A gene product required for correct initiation of segmental determination in *Drosophila. Nature* 293:36–41.

Tamm, S. 1978. Relations between membrane movements and cytoplasmic structures during rotational motility of a termite flagellate. Abstracts, Cold Spring Harbor Cytoskeleton Meetings, p. 89.

Teilhard de Chardin, P. 1920. Le cas de l'homme de Piltdown. *Revues des questions scientifiques* 27:149–55.

Teilhard de Chardin, P. 1955– . Complete edition of letters and general works (in French). Paris: Editions de Seuil (13 volumes so far).

Teilhard de Chardin, P. 1959. *The phenomenon of man.* New York: Harper and Brothers.

Teilhard de Chardin, P. 1965. *Lettres d'Hastings et de Paris, 1908–1914.* Paris: Aubier, Editions Montaigne.

Thompson, D. W. 1942. *Growth and form.* Cambridge: Cambridge University Press.

Thorpe, W. H. 1956. *Learning and instinct in animals.* Cambridge, MA: Harvard University Press.

Trueman, A. E. 1922. On the use of *Gryphaea* in the correlation of the Lower Lias. *Geological Magazine* 59: 256–68.

Vavilov, N. I. 1922. The law of homologous series in variation. *Journal of Genetics* 12:47–89.

Weiner, J. S. 1955. *The Piltdown forgery.* London: Oxford University Press.

Winchell, A. 1870. *Sketches of creation.* New York: Harper and Brothers.

Wood, A. E. 1957. What, if anything, is a rabbit? *Evolution* 11: 417–25.

Index

BY THE SAME AUTHOR

Time's Arrow, Time's Cycle

'Gould provides a fascinating, informally written excursion into the ways we conceptualize the past. Entertaining, sometimes annoying, highly personal, but never dull, this is the shortest of Gould's books, but also his most adventurous and experimental' – *The Times Higher Education Supplement*

The Mismeasure of Man

Can human intelligence be measured? Elegantly written and tightly argued, this book exposes the fatal flaws in intelligence testing.

'Superlative' – *Nature*

The Panda's Thumb

'From the opening essay on the panda's thumb to an intriguing vignette on the Adactylithum mite which dies before it is born, the book provides a quirky and provocative exploration of the nature of evolution . . . wonderfully entertaining' – *Sunday Telegraph*

Ever Since Darwin

'A stimulating tour of new ideas in palaeontology, botany, geology and much else – a marvellous read' – *British Medical Journal*

An Urchin in the Storm

'A new collection of superb essays by this distinguished scientist is always a cause for rejoicing . . . There is not a dull or uninformative paragraph in the book' – Martin Gardner